# The ZX Spectrum ULA
## How to Design a Microcomputer

**Chris Smith**

chris@zxdesign.info

**ZX Design and Media**

**The ZX Spectrum ULA**: *How to Design a Microcomputer*
by Chris Smith

Published by ZX Design and Media

23 Stacey Road, Dinas Powys
Vale of Glamorgan. CF64 4AE
United Kingdom
info@zxdesign.info
http://www.zxdesign.info


First Edition
Published 2010
Copyright © 2010 Chris Smith
ISBN-13: 978-0-9565071-0-5

Edited by Andrew Owen

This book is printed on acid-free paper.

# Dedication

Dedicated to my wonderful wife Marcia, without whose patience and support the huge undertaking that became the writing of this book would never have got further than a collection of notes.

# Contents

# List of Figures

# Preface

At the age of 11 in the early 1980s I began exploring the world of micro-computers, first with the Sinclair ZX81, and then with the ZX Spectrum. I marvelled at the seemingly endless possibilities they offered, and was in awe of the engineers who conceived, designed and built the machines.

From the moment I acquired my first microcomputer, I started designing and building electronic add-ons for them. My ambition however, was to build my own computer.

The biggest obstacle to realising my dream was designing the television or monitor interface. I bought and borrowed many electronic and microelectronic books, trying to cross-reference and compile enough information to learn the concepts and techniques I needed to master. However, the information I found was generally very high level and vague, or written for the professional electronics engineer and therefore almost impossible to decipher by the untrained eye. There simply were no books that dealt with the subject of building your own microcomputer, or the techniques involved in graphical display generation, and so I could not achieve my goal.

Many years later, in 2007, I uncovered some of my notes that had lain hidden for 20 years, and was inspired to revisit my childhood dream and apply the benefit of further education and experience. To that end I decided to design and build my own the ZX Spectrum from scratch, aiming for 100 per cent compatibility. In the course of my work, I made several previously undocumented discoveries about the ZX Spectrum's design, and was encouraged by my friends and enthusiasts of the ZX Spectrum to document my findings in a short book.

In writing a book, I wanted to ensure that every detail was factual, and not merely inferred through non-invasive reverse engineering and experimentation. Achieving this was going to be a major hurdle, as 25 years had passed since the ZX Spectrum was designed, and there was little hope of obtaining the original circuit diagrams.

My luck changed in November 2008 when Mike Connors of Datel Ltd offered to optically image the silicon chip of the ULA. With these images and the help of the Ferranti archive at the Museum Of Science and Industry in Manchester, I was able to work out the component structure of the ULA and back-annotate the entire ZX Spectrum ULA schematic.

To do justice to the wealth and quality of information I now had, I felt compelled to write a bigger, definitive guide to the ZX Spectrum ULA, and perhaps the book I had sought during the 1980s.

The back-annotation of the ULA images into a full schematic took approximately one year to complete. I first produced a paper schematic that matched the physical layout of an uncommitted 6000 series logic array, and copied onto this the connections from between the 2,500 transistors in the ULA image.

From this transistor-level schematic I went on to produce a NOR gate schematic of the same size and structure. Within it I outlined individual functions such as flip-flops and latches, and grouped them into functional units such as counters, identifying and labeling signals as I proceeded. Once this schematic was complete, functional analysis was possible.

The most difficult chapter to write has been Chapter 13, *Video Memory Access*. The circuit itself is functionally quite simple, however the signal timings measured at the ULA pins do not match those suggested by the circuit. The parameters of the 6C001 logic gates are currently unknown, and this makes it difficult to predict the timing of signals that are intentionally subjected to propagation delay. Because of this, where discrepancies exist due to insufficient information being available, a comment has been made to this effect.

The most rewarding chapters to research and write have been Chapter 5, *The Ferranti ULA*, Chapter 6, *Sinclair and the ULA*, Chapter 16, *Analogue Video* and Chapter 23, *Hidden Features and Errors*. Each chapter offered its own challenges, and made them all the more enjoyable to work on.

I hope the reader finds the information presented in this book informative and useful, and that it answers the questions they have about microcomputer design and the Sinclair ZX Spectrum ULA.

# Acknowledgements

Several people have been instrumental in the production of this book.

I would like to thank Richard Altwasser for helping to fill in the blanks and providing a real sense of what it was like to be a young engineer at Sinclair Research, working to such tight cost and time constraints.

I would like to thank Mike Connors at Datel Limited who kindly decapsulated and optically imaged a number of ULAs, a highly skilled and laborious process that would have been prohibitively expensive for me to undertake without Mike and his team's kind assistance. Having the ULA image made the back-annotation of the full ULA schematic possible, enabling the correction and verification of the schematics and material presented in this book.

I would like to thank Jan Shearsmith, archivist at the Manchester Museum Of Science and Industry, for allowing access to their Ferranti archives and for permission to use photographic material in this book. Without his assistance the description of the Ferranti ULA, the technology and company behind it would have otherwise been inaccurate and incomplete.

Thanks to Dylan Smith and Mike Hally for their last minute proof reading and corrections.

Thanks also to Paul Hartley for kindly donating his copy of the ULA Technical Handbook, a valuable source of information.

Finally I would like to thank Andrew Owen for his consistent editorial advice and for encouraging me to use Docbook XML. I would also like to thank the members of the comp.sys.sinclair usenet news group and World Of Spectrum web forum for their encouragement and patience while I was writing this book.

Acknowledgements

# Chapter 1
## Introduction

The objective of this book is to present the design and implementation of the Sinclair ZX Spectrum's core and custom chip, the Uncommitted Logic Array (ULA), and so doing introduce the concepts and methodologies required to design a microcomputer based around an 8-bit microprocessor.

The ULA contains the video display logic responsible for converting data stored in video RAM into a television picture. It controls the CPU and manages its access to memory, along with decoding the keyboard and cassette interfaces. In effect, the ULA glues the CPU, RAM and ROM together so that they form a ZX Spectrum, and as such it is the soul of the machine.

The book is intended for two different audiences:

- The electronics hobbyist or professional researching the designs and techniques pioneered and used by home computers of the 1980s, either wishing to design their own computer or for purely academic reasons.
- The ZX Spectrum enthusiast wishing to discover the secrets that the ZX Spectrum ULA holds.

Material is presented in a form that is accessible to the amateur, but thorough enough for the expert. Some familiarity with basic circuit design, digital electronics and boolean algebra is assumed.

Accurate schematics are given for the complete ZX Spectrum ULA and are accompanied by comprehensive analysis and discussion. With these schematics it is possible to implement a 100 per cent compatible clone of the 48K ZX Spectrum using discrete or programmable logic, such as a CPLD.

In addition, for the first time, a full discussion of the Ferranti ULA technology and manufacturing process is given, a technology without which Sinclair

Research and others could not have brought home computing within reach of the general public.

To present the details of the ZX Spectrum ULA, its design has been divided into logical sections and a chapter assigned to each. On the whole, a chapter builds upon the material presented in previous chapters, and where possible cross-references have been provided to assist the reader who is using the book as a reference text.

The operation of the ZX Spectrum's PCB circuit is not discussed, except where specific details are relevant to the ULA design. Circuit schematics for the ZX Spectrum PCB are freely available via the Internet.

Ferranti's ULA technology occupies a unique place in the development of the microelectronics industry. It relied on an advanced transistor fabrication method called Collector Diffusion-Isolation (CDI). To understand what made the ULA significant, an understanding of CDI and how it differs from other fabrication methods is required. A reader familiar with such semiconductor transistor fabrication techniques may wish to skip the overview provided by Chapter 2, *Integrated Circuits*.

Chapter 3, *The Standard Microcomputer*, introduces the standard design on which all computers of the late 1970s and beyond were based, including the modern PC. It also presents an overview of the Z80 microprocessor, around which the ZX Spectrum was designed.

Chapter 4, *Semi-Custom Devices*, discusses advances in microelectronic fabrication that allowed manufacturers to realise the potential of custom designed integrated circuits, without resorting to the prohibitively expensive custom made silicon chip.

Chapter 5, *The Ferranti ULA*, introduces Ferranti Semiconductors, their CDI process and range of ULA products. The ULA customisation process is then described in detail, starting with the customer specification and moving through design, fabrication and testing.

Chapter 6, *Sinclair and the ULA*, gives a brief history of the development of the ZX80 and ZX81, predecessors of the ZX Spectrum, and discusses Sinclair's first use of the Ferranti ULA. This is followed by a description of the development process and testing of the ZX Spectrum. Although the MK14 was technically Sinclair's first computer, it does not fit the category of a home microcomputer, and therefore it is not discussed here.

Chapter 7, *The ZX Spectrum Overview*, introduces the functional areas of the ZX Spectrum's design, and the relationship between them. Those read-ers familiar with these may wish to skip this chapter.

Chapter 8, *The Memory Map*, describes how the memory of the ZX Spectrum is allocated to the system, display and user.

Chapter 9, *The Video Display*, introduces the principles of television display generation, and describes the concepts required to design a computer display.

Chapter 10, *The Internal Clocks*, discusses the various clocks and counters required to track the position of the television's electron beam as it scans the screen. These counters are at the core of the ZX Spectrum's ULA design, and control all aspects of its operation.

Chapter 11, *Video Synchronisation*, explains how the television electron beam is synchronised with the counters described in Chapter 10, *The Internal Clocks*, allowing a stable and flicker-free picture to be displayed.

Chapter 12, *Generating The Display*, describes how a computer display is designed, what resolution it should have, and how the digital information that is stored in memory can be processed into a form suitable for conversion into a video signal.

Chapter 13, *Video Memory Access*, discusses the control signals that the video generator must produce to enable it to access the computer's video memory. Specific reference to the ZX Spectrum's signal timing is given.

Chapter 14, *Video Control Clocks*, gives a full analysis and design of the video control signals identified in the preceding chapters.

Chapter 15, *Video Addressing*, introduces the memory addressing schemes suitable for use with the video system discussed so far, both theoretical and practical. In doing so, the ZX Spectrum's peculiar video memory arrangement is explained.

Chapter 16, *Analogue Video*, gives a thorough discussion and analysis of the techniques required to generate a video signal composed of luminance and chrominance signals. The design of the ZX Spectrum's analogue video output is covered in detail.

Chapter 17, *CPU Memory Access*, discusses the control signals that must be generated to enable the CPU to access the memory within the computer.

Chapter 18, *CPU Clock and Contention*, describes the conflict that exists between the CPU and the video generator when they simultaneously require access to the video RAM. Several conflict resolution methods are discussed, culminating in the explanation of the techniques used by the ZX Spectrum, including the differences between the three issues of ULA.

Chapter 19, *Input-Output Devices*, discusses the input and output (I/O) capabilities of the ZX Spectrum and the design behind these interfaces, making reference to Chapter 3, *The Standard Microcomputer*.

Chapter 20, *Cassette Storage and Sound*, discusses the analogue requirements of the ZX Spectrum's cassette and loudspeaker interfaces, and describes their patented design.

Chapter 21, *Interrupts*, follows on from the topic introduced in Chapter 3, *The Standard Microcomputer*, and analyses the interrupt timing implemented in the ZX Spectrum ULA. The cause of the "late timing" interrupt detection is also discussed.

Chapter 22, *Signal Interfacing*, gives the complete list of signals passed in and out of the ZX Spectrum ULA, describing their purpose, signal level and capability.

Chapter 23, *Hidden Features and Errors*, uncovers undocumented errors and hidden features of ZX Spectrum ULA design, and examines the cause and effect of design and production errors such as the "snow effect".

Chapter 24, *ULA Versions*, gives a complete overview of each ZX Spectrum ULA revision, highlighting the significant differences between them.

Appendix A, *The ULA Die Plot*, illustrates the physical layout of the ZX Spectrum ULA's silicon die at a functional level, and relates each area to the circuit designs presented in this book.

Appendix B, *Component Library*, gives the basic circuit building blocks used in the design of the ZX Spectrum ULA. All of the complex components used, but not described elsewhere in this book (such as flip-flops, shift registers and TTL outputs), are given here.

Appendix C, *ULA Configuration*, explains how ULA matrix cells are interconnected and intraconnected to provide different logic gate configurations and speeds. This chapter supplements Chapter 5, *The Ferranti ULA*.

A glossary of terms is provided at the end of the book, along with a comprehensive bibliography.

# Chapter 2
# Integrated Circuits

Integrated circuits became possible with the invention of the Planar process in 1959 by Jean Hoerni at Fairchild [HOERNI]. This viewed the silicon wafer substrate as a flat two-dimensional plane onto which successive layers of silicon could be deposited through oxide masks and Photolithography, allowing all components of a transistor to be fabricated from one side of the silicon wafer. Shortly afterwards in July of that year Robert N. Noyce, also at Fairchild, devised the monolithic integrated circuit [NOYCE] where several transistors are formed and interconnected on a single silicon die, and produced the first working ICs in May 1960. Publicly available devices were announced in March 1961, photographs of which appeared in LIFE magazine. It is generally recognised that Jack Kilby of Texas Instruments independently conceived the monolithic integrated circuit at about the same time, and though both companies fought legal battles over patents, they eventually settled and cross-licensed their technologies.

In 1960, Bell Labs introduced a process of growing a thin layer of silicon on the substrate by chemical-vapor deposition to provide isolation between a transistor's base and collector regions. This epitaxial process reduced current leakage, increased the breakdown voltage and dramatically increased the switching speed of the transistor. In 1961, Jean Hoerni increased the switching speed of silicon transistors further, to exceed that of germanium, by doping silicon with gold impurities. Silicon is preferable to germanium as it has a wider temperature tolerance and is stable at up to 150°C, twice that of germanium. This allowed the first high speed computers to be produced, incorporating many hundreds of transistors that generate considerable amounts of heat. Germanium transistors do not switch reliably at the temperatures found within these machines, making them unsuitable.

The UK-based Ferranti Semiconductors began experimenting with monolithic ICs around 1961, having successfully marketed a range of diffused transistors such as the 1960 ZT20 Mesa transistor. They introduced the Micronor I IC in

1963/64, Europe's first integrated circuit, followed in 1965 by the Micronor II, both Diode-Transistor Logic (DTL) devices [SWANN].

In addition to bipolar transistor technologies, Fairchild was also developing the Complimentary Metal Oxide Semiconductor transistor, which was first described by Sah and Wanlass in 1963. CMOS combines n-channel and p-channel MOS transistors in a complementary symmetry configuration, which draws almost zero power in standby mode. These early CMOS devices were however plagued by manufacturing and reliability issues, and it would be several years before they saw high volume adoption, eventually becoming the dominant technology of modern integrated circuits.

It was during 1963 that the first mass market for digital ICs began, initially based around Signetics SE100 Series Diode-Transistor Logic, followed by Farichild's cheaper and better performing 930 Series. At the same time, Sylvania introduced its SUHL Transistor-Transistor Logic (TTL) series, the success of which encouraged Texas Instruments to introduce its SN5400 TTL Series the following year. In 1966 TI announced the SN7400 series and quickly gained more than 50% of the market.

The significant weight and size reduction of integrated circuits compared to discrete transistors prompted the Massachusetts Institute of Technology and NASA to select a Fairchild Micrologic 3-input Resistor-Transistor NOR gate for the Apollo Guidance Computer, becoming the largest single user of ICs in 1965.

Driven by the demands of high power mainframe computing, circuit configurations offering significant improvement in speed and performance were sought. To this end, mainframe manufacturer RCA developed the concept of Current-Mode Logic (CML) circuits, custom produced for their Spectra 70 series computer by various IC manufacturers (1965). Fairchild developed a similar Complementary Transistor Logic (CTL) family to power the Burroughs B2500/3500 and Hewlett-Packard's 3000 Series (1966).

## Epitaxial Transistor Fabrication

There are a variety of methods of fabricating bipolar transistors, and we shall consider those used in IC manufacture and which specifically use the epitaxial process.

Bipolar transistors use both hole and charge carriers, and their fabrication is a planar process where regions of an n-type or p-type silicon substrate are doped using high temperature gas diffusion. The areas to be doped are typically made through windows in a silicon dioxide mask, and depending on the doping element used, n-type or p-type areas are formed. Where two different types of silicon are in contact, a **PN Junction** is created which acts as a conductor when voltage is applied in one direction, and a nonconductor in the other, as in Figure 2-1.

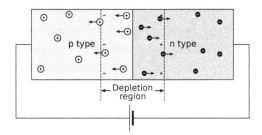

*Figure 2-1: PN junction showing hole and charge carriers*

Doping an n-type area with a group 13 element such as boron produces a p-type area of electron acceptors (also known as holes). Doping a p-type area with a group 15 element such as arsenic, produces an n-type area of electron donors.

## The Standard Buried Collector Process

The general transistor fabrication of an NPN transistor using the Standard Buried Collector (SBC) process is as follows (Figure 2-2):

1. An area of highly doped n+ is diffused into a p-type substrate. This is referred to as the buried sub-collector.
2. An epitaxial layer of n-type silicon is grown over the substrate.
3. A deep diffusion of p+ is made through the epitaxial layer, creating a deep moat around the device to form an isolated island.
4. A shallow diffusion of lightly doped p-type silicon is made, creating the base region.
5. A shallow diffusion of n+ is made in the base region forming the emitter, and in the epitaxial layer forming the collector.
6. At the emitter, base and collector regions, aluminium is deposited to create the transistor contacts.

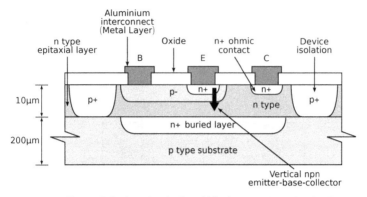

*Figure 2-2: Junction-isolated bipolar transistor fabrication*

The epitaxial layer and diffusions are thin, so the NPN transistor currents flow vertically from the emitter, through the base to the collector.

Isolation between multiple devices on the same silicon die is achieved by making the p-type substrate the most negative point. A PN junction therefore exists between the p-type moat and the n-type collector region, and will be reverse-biased, creating what is termed junction or diode isolation.

The n+ buried layer prevents the formation of a vertical parasitic PNP transistor between the base, collector and substrate. The shallow diffusion of the n+ ohmic collector contact is required as aluminium and n-type silicon will together form a slight PN junction. Buffering the aluminium contact with a highly doped n+ region prevents such a junction forming.

SBC transistors do not make efficient use of the silicon die as the active area of the SBC transistor is only in the region directly below the emitter. Also, due to lateral diffusion the minimum width of the p+ isolation moat is twice the depth of the epitaxial layer, so that the useful area of the transistor is less than 5% of the total device area, with the active area beneath the emitter being only 2.67% of the total device [HURSTVLSI].

Integrated circuits are generally constructed from NPN transistors, as the alternative PNP transistor has a lower performance, with holes, not electrons, forming the majority charge carrier. Also, the transistor action tends to act horizontally in SBC PNP transistors.

Other components such as diodes and resistors are created by a similar process of isolating epitaxial regions of n-type silicon with p+ diffusions.

### Standard Resistor Fabrication

Resistors are formed by surrounding a region of epitaxial layer above a buried p+ layer, with a deep diffusion of p+. This creates an isolated region of n-type silicon with a characteristic sheet resistance. Metal deposits at either end of the region provide the resistor contacts.

By varying the width of the n-type silicon region and the distance between the contacts, different resistor values can be formed. The sheet resistance of an epitaxial layer is usually given in units of ohms per square, where one square is the distance between the resistor contacts divided by the width of the region.

1. An area of highly doped n+ is diffused into a p-type substrate. This is referred to as the buried layer.
2. An epitaxial layer n-type silicon is grown over the substrate.
3. A deep diffusion of p+ is made through the epitaxial layer, creating a rectangular boundary around the resistor, creating an isolated region of n-type silicon.
4. Aluminium deposits are placed at either end of the longest resistor dimension, forming the resistor contacts.

## Logic Gate Technology

A single bipolar transistor provides only the simplest logic function, that of the inverter. To build more complex logic functions, transistors must be interconnected to form functionally complete[1] gates which can be built up into higher functional units.

Various families of bipolar logic have been used by manufacturers to produce a range of off-the-shelf logic devices, with each family having particular useful characteristics.

### Resistor-Transistor Logic

Resistor-Transistor Logic (RTL) is one of the simplest and earliest class of logic gate, and was used to create the first integrated circuits in March 1961. A typical three input NOR gate is shown in Figure 2-3.

The disadvantage of RTL devices is their high current consumption and slow switching speed due to transistor saturation and charge storage. This high power requirement leads to issues with heat dissipation, which limits the gate packing density and therefore the number of gates per chip.

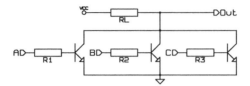

Figure 2-3: Three input RTL NOR gate

## Transistor-Transistor Logic

This is the standard and most popular bipolar logic family used in small and medium scale integrated circuits. Figure 2-4 shows a typical TTL three input NAND gate with a totem pole output, giving active pull up to Vcc for logic 1 and push down to 0v for logic 0.

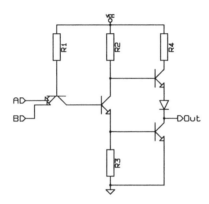

Figure 2-4: TTL NAND gate with totem pole output

The disadvantage with this logic family is it's high power consumption, slow propagation times caused by transistor saturation, and large current spikes when the output switches between logic levels due to a moment when both transistors conduct, shorting Vcc to 0v.

To address these problems several advances in TTL design have occurred, most importantly the Schottky TTL gate. Here a Schottky diode is placed between the base and collector of the switching transistors to prevent them becoming saturated and storing charge, greatly improving their switching speed. Figure 2-5.

A Schottky diode is formed between lightly doped n-type silicon and aluminium, which may be considered a weak p-type dopant, thus forming a PN junction. However, as few holes are produced by the weak p-type aluminium, most of the semiconductor action consists of electron donors which gives rise to junction characteristics that are different from normal PN junctions; for instance, a forward voltage drop of approximately 0.35v and practically zero storage time. A Schottky diode placed across the transistors base-collector junction will therefore conduct before the transistor (at 0.35v compared with 0.6v), redirecting current and preventing transistor saturation.

*Figure 2-5: Schottky diode and transistor*

Fabrication of a Schottky diode across the collector-base junction is easily achieved by extending the aluminium that forms the ohmic contact with the p-type region of the base, so that it overlaps the area of n-type silicon at the collector, where it reaches the transistor surface. Figure 2-6.

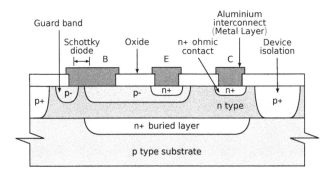

*Figure 2-6: Schottky TTL fabrication*

11

**Emitter-Coupled Logic**

Emitter-Coupled Logic (ECL) is one of the fastest switching class of logic gate, at the expense of high power dissipation which limits the number of gates per device. They have generally been used for the highest performance computers, such as the CRAY-1.

ECL gates work by directing a fixed current through one of two paths, depending on the input logic levels. In this way a current always flows, removing voltages spikes during output switching but contributing to the overall power consumption. Figure 2-7 shows a simple ECL NOR gate. Multi-input NOR/OR gates are created by connecting multiple input transistors in parallel. With a Vin less than VRef, T1 will be off and T2 will be conducting. As Vin exceeds VRef, T1 begins to conduct and T2 shuts off. Two complementary outputs are produced by the two switching transistors, allowing both NOR and OR functionality to be simultaneously implemented. The resistors are chosen so neither transistor saturates when it conducts, achieving the maximum switching speed. The low output voltage swing produced as a result complicates external interfacing, where 5v signals are normal.

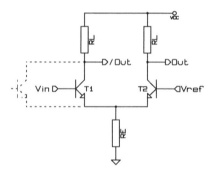

*Figure 2-7: Typical ECL gate*

**Current-Mode Logic**

Unlike ECL, where a constant current is switched between one of two paths, Current-Mode Logic (CML) allows a preset current to flow or not to flow through the switching transistors, depending on the state of the logic input. Its basic circuit configuration is shown in Figure 2-8.

The value of the preset current supplied by the current source is small enough to prevent the transistors saturating, so that they only operate in their off or active mode, thus enabling them to switch rapidly.

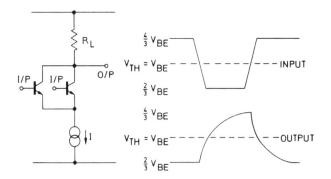

Figure 2-8: CML NOR gate and signal thresholds

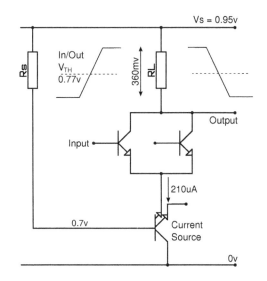

Figure 2-9: CML NOR gate showing output voltage swing

CML gates may be produced in a wide range of speed and power configurations, as their speed power product is directly proportional to the supply voltage and logic swing; therefore by reducing the size of the load resistor, RL, or increasing the supply voltage, the gate switching speed may be increased at the expense of increased power dissipation.

Typical CML gates have a minimum supply voltage of 0.6 to 0.95 volts, and a logic swing range which is determined by noise considerations. Under LSI chip conditions, where noise is controlled, a margin of 200mV is equivalent to a 1V noise margin found within a printed circuit board; therefore a logic swing of approximately 360mV within the gate is more than sufficient (Figure 2-9), and is defined by the current source and load resistors.

The low on-chip voltage swings achieved, however, means that signals require amplification and buffering before they can be connected to external devices.

The constant current source in the emitter circuit is provided by a multi-emitter transistor operating in inverse mode. That is, its collector is used as the emitter and its emitters as multiple collectors, one of which has a particular base-collector bias that causes it to act as a current mirror, setting the current that will be passed by the other collectors. See Figure 2-9.

This current source action demands a transistor that exhibits extremely good inverse-mode operation, something that Collector-Diffusion Isolation (CDI) fabrication in particular provides.

## Collector-Diffusion Isolation Process

Collector-Diffusion Isolation (CDI) process was invented by B. T. Murphy et al at Bell Labs in 1969 [MURPHY]. The technology proved troublesome for Fairchild and Bell Labs who, to date, had only managed to make Collector-Diffusion Isolation integrated circuits that operated at 3 volts, and not at the industry standard of 5 volts. Ferranti Electronics Limited licensed and further developed this technology for mixed digital and analogue applications, becoming the worlds first microelectronics supplier to successfully utilise CDI in the production of VLSI integrated circuits [WILSON2].

The CDI process is similar to that of the SBC, except that a p-type epitaxial layer is grown instead of an n-type layer, eventually becoming the final transistor base regions. The CDI process is as follows:

1. A low-resistance n+ buried layer is diffused into the p-type substrate.
2. A p-type epitaxial layer is grown over the substrate, eventually forming the base region of the transistor.

3. A deep n+ isolation moat is diffused through the epitaxial layer forming an isolated region of p-type silicon. This n+ diffusion also forms the collector of the transistor, surrounding the base region.

4. A p+ diffusion is made to create a shallow p+ skin over the whole silicon slice. The concentration of this diffusion is so low that it does not alter the polarity or resistivity of the n+ isolation diffusion. It does however perform the important task of ensuring that no inversion occurs at the surface, and creates a sheet resistance of 470 ohms per square for fabricating resistors.

5. A shallow n-type diffusion into the epitaxial layer forms the emitter. This pushes the p+ skin diffusion down into the epitaxial layer creating a graded base transistor with controlled current gains and a high gain bandwidth product.

6. At the emitter, base and collector regions, aluminium is deposited to create the transistor contacts.

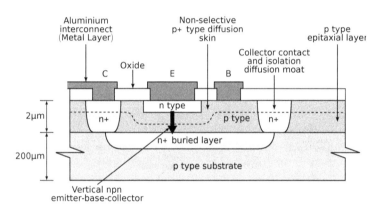

*Figure 2-10: Collector diffusion-isolation fabrication*

The CDI process is far simpler than the SBC process, and provides many benefits. In particular it has a thin epitaxial layer, of the order of $2\mu$m, and very shallow diffusion depths of approximately $1\mu$m, creating a very narrow base width and good operating speed. Silicon is therefore used economically, achieving circuit packing densities 2.5 times higher than obtainable using the standard buried collector structure with the same masking tolerances [ULA-HAND][MURPHY].

Another feature of the CDI process is that the heavily doped n-type collector region gives the transistor a good inverse-mode operation, required for current sources in Current Mode Logic circuits. In addition to allowing the p-type substrate to be used as the ground connection, CDI devices allow n+ diffusions to be used to distribute the supply rail around the chip.

### CDI process resistor fabrication

As with the SBC process, resistors are formed by surrounding a region of epitaxial layer by a deep moat of diffused n+, over a buried n+ layer. This creates a completely isolated region of p-type silicon which, due to its diffused p+ skin, has a defined sheet resistance. Metal deposits at either end of the region provide the resistor contacts.

Altering the length and width dimensions of the resistor alters the value of the resistor.

1. An area of highly doped n+ is diffused into a p-type substrate.
2. An epitaxial layer, or skin, of p-type silicon is grown over the substrate.
3. A deep diffusion of n+ is made through the epitaxial layer, creating a rectangular boundary around the resistor which connects with the n+ buried layer, creating an isolated region of n-type silicon.
4. Aluminium deposits are placed at either end of the longest resistor dimension, forming the resistor contacts.

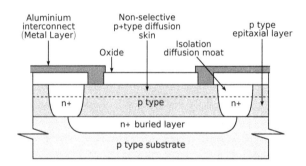

*Figure 2-11: CDI process resistor fabrication*

---

1. A functionally complete gate is one that may be used as the building block of any other logic gate. NOR and NAND are such gates.

# Chapter 3
# The Standard Microcomputer

The standard microcomputer of the late 1970s and early 1980s consisted of four separate components: a microprocessing unit (MPU or CPU), Read Only Memory (ROM) for program storage, Random Access Memory (RAM) for data and program storage and some form of input and output (IO). These components were produced as standalone devices, designed to be integrated into such a system architecture, as the degree of fabrication density possible at the time did not permit integration of memory and I/O directly into the microprocessor chip itself, what would later become termed the System-on-a-Chip (SoC).

All microcomputers of the era followed this design, using common off the shelf memory ICs and peripheral components. It was only the choice of microprocessor and custom I/O circuits that stood them apart.

## The Architecture

The microprocessor prevalent at the this time was the 8-bit processor. Such devices could transfer eight data bits at a time and were usually provided with up to a 16 bit address bus, enabling access to 65536 ($2^{16}$) memory locations. Control signals were also necessary to coordinate the activity of devices connected to the processor, up to 14 lines in total, as well as the power supply. Typically as many as 40 pins would be used. Producing ICs with more than 40 pins was prohibitively expensive due to their size, and this limited the number of CPU data and address lines that could be made available to memory and I/O devices. The few 16-bit devices that were produced at the time provided the additional data and address bus signals by multiplexing them over existing signals, which increased the complexity of the system design greatly, and was generally considered not worth the effort.

The basic architecture of the standard microcomputer is shown in Figure 3-1. To the far left, a clock circuit provides the CPU with a regular time signal with which to synchronises its internal state machine. Next is the CPU itself, providing address, data and control buses through which it interfaces with ROM, RAM and I/O devices, shown to the right.

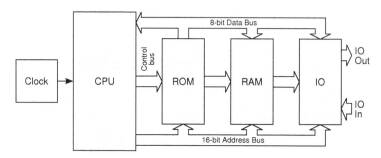

*Figure 3-1: The standard microcomputer*

## The Control Bus

The control bus contains signals that allow the processor to control the operation of devices to which it is connected, and allows devices to control certain aspects of the processors operation. In general the signals provide the following activities:

1. To notify memory devices that the processor is about to access a memory location and that they should use the address on the bus to determine which location is required.
2. To notify I/O devices that the processor is about to access an I/O port and that they should use the address on the bus to determine which port is required.
3. To interrupt the processor and have it perform a time critical task.
4. To pause the processor, allowing slow memories or other devices time to prepare for the requested memory or I/O operation.
5. To request and acknowledge a direct memory access (DMA).
6. To specify the direction of data flow.
7. To reset the processor.

The exact nature of these control signals, and which are provided, depends on the microprocessor. For example, some processors provide signals to select memory or I/O devices along with read and write signals; whereas others make I/O devices synonymous with memory and provide just read and write signals.

A typical memory write by the processor would see it place the address of the required memory location on the address bus and activate the memory access signal. While the memory device is responding to this signal, the processor places the data to be written on the data bus and activates the write control signal. The memory detects this write indication and stores the data in the pre-prepared location. The processor then deactivates the two control signals.

## Memory Devices

Memory ICs may be either ROM or RAM and generally provide fewer storage locations than the 65536 that may be addressed by a 16-bit address bus. Several devices will therefore be necessary to provide both ROM and RAM at the quantity of memory required. These devices must be connected to the CPU in such a way that they appear at their appropriate place in the *memory map* and provide contiguous pools of memory locations.

The placement of devices within the memory map is achieved through address bus decoding, which may take the form of linear(partial) or full address decoding.

Linear decoding works by assigning the more significant signals of the address bus to the select pins of individual memory devices. This simple decoding results in a fragmented memory map because devices are mapped to blocks of address that increase in powers of two. This type of address decoding is therefore not suitable when several memory devices must be placed one after another in the memory map.

Full address decoding considers combinations of address bus signals and activates specific memory devices when certain ranges of address appear on the bus. This decoding is more complicated than linear addressing and requires additional logic, but it does allow devices to be placed at any location within the memory map.

### Dynamic RAM

The memory capacities prevalent in the 1970s and 1980s were small by today's standards, but back then silicon die sizes were very much larger, and fabrication of as little as 32 or 64K RAM was expensive - comparable in price to the multi-megabyte devices available today.

There are two types of RAM generally used in computers: dynamic RAM and static RAM. Dynamic RAM employs a capacitor charge method of storing the binary value of bytes. However this state leaks away as the capacitors discharge, so it must be regularly refreshed by external control circuitry. Static RAM by contrast uses flip-flops to store binary values, which do not forget their state. Flip-Flops are relatively complicated and therefore take up more space on a silicon wafer than a capacitor, consume a larger current and are more expensive to produce. Dynamic RAM (DRAM) therefore offers a higher density and cheaper overall cost than static RAM (SRAM), even with the added complexity of refresh control.

Internally RAM is arranged as a matrix, or grid, of storage locations. To access any one location, its row and column must be selected, the location being accessed being at the intersection of the two.

To reduce cost further, the physical package size of a RAM device is kept to a minimum, the limiting factor being the number of connection pins required. DRAM designers reduce the number of pins by dividing the address bus in two, and have both halves share the same address pins. This makes interfacing more complicated as the CPU address bus must be presented to the DRAM in two stages.

This technique is call multiplexing, and DRAM chips provide two special pins to achieve this: a Row Address Strobe (RAS) and a Column Address Strobe (CAS).

Systems using DRAM place the lower half of the memory address on the DRAM address bus and pull RAS low, causing the DRAM to select the corresponding matrix row. Shortly after this the upper half of the memory address is presented to the DRAM on the same bus, and the CAS signal pulled low, selecting the required matrix column and thus the desired memory location.

In addition to selecting a memory location, the state of the RD/WR signal determines whether the chosen location is to be read from or written to.

To keep the contents of the dynamic RAM refreshed and prevent the stored charge leaking away, each matrix row must be read every 2 milliseconds (ms) or less. This is performed by initiating a RAS-only refresh cycle by placing a refresh row address on the RAM data bus, and activating the RAS control line. The refresh row address is then incremented and used for the next refresh cycle, 2ms later.

### ROM

The interfacing requirements of ROM are simpler than for dynamic RAM because they are non volatile devices that do not forget their contents when power is removed, and do not require refreshing. They are programmed once when they are manufactured, and their contents cannot be changed.

## Input and Output

A computer essentially takes a number of inputs and processes them to produce one or more outputs. To be useful, the standard microcomputer described above must be provided with both input and output devices. Output devices may consist of LEDs, seven-segment displays, a CRT monitor, digital to analogue converters (DAC), printers or tape recorders. Inputs devices include switches, keyboards, analogue to digital converters (ADC) and tape recorders.

Interfacing such I/O peripherals to the microprocessor can be achieved in two ways. One, through memory mapping the device, or two, by I/O mapping the device, if supported by the CPU.

Memory mapping input–output devices may be a matter of choice, or the only option if the microprocessor does not provide I/O specific control signals and instructions. Here the I/O device is allocated a region of the memory map through memory address decoding. This has the advantage of allowing input and output to be performed through the processor's wide range of powerful memory instructions. The disadvantage being that memory mapped I/O reduces the amount of system memory that can be provided.

For I/O mapped input–output, the processor provides a set of I/O specific enable and direction control signals, complementing those for memory devices. It also provides a set of I/O related instructions through which input and output may be performed. The advantages of this scheme are: One, the full address space is available for memory devices. Two, the processor provides instruction timings and features specifically for I/O. Three, program code that accesses an I/O device is easily distinguishable from that which accesses memory. The disadvantages are the loss of the powerful memory instructions and the need for the processor to provide additional control signal pins.

### Serial I/O

Many peripherals of the era communicated with the microcomputer via a serial I/O interface. Data transfer rates were slow, around 300 bits per second typically, but required only a single connection wire for each direction of data flow and another for synchronisation.

The basis of serial communication is to convert each byte to be sent into a stream of 1s and 0s, and to reassemble a stream being received back into bytes. This can be achieved in two ways: through software or through a dedicated I/O communications IC called a universal asynchronous receiver-transmitter or UART.

With software, a program takes responsibility for the serialisation and deserialisation of data. For output, the program takes each byte in turn and shifts bits out of it with one of the processor's shift instructions, at a fixed rate. For each bit examined, a single bit of an output port is set or reset. Where handshaking is required, an additional output port bit is set and reset as required. For input, the program samples the input port bit until it detects a start bit, after which it samples the port at the required rate, shifting the bits read into the receiving byte with another of the processor's shift instructions.

The advantage of software serialisation and deserialisation is its flexibility and the use of specialised UART hardware is avoided. The disadvantages are: one, the software can be complicated and ties up the processor, preventing it from carrying out other tasks, and two, it is slow.

Hardware UART ICs take the burden of converting bytes to and from a serial data stream away from the processor, freeing it to perform other tasks like processing the data that is being received or handling other events. It is interfaced to the microprocessor through two I/O ports: one for bi-directional data input and output, and another for interface control. UARTs simplify software design, lighten processor load and increase data throughput, however this comes with increased hardware complexity and cost.

**Keyboard Input**

Keyboards may be interfaced to the microcomputer in two ways: One, as a serial peripheral that sends information about key presses over a serial data connection. Two, as a bank of switches connected to the microprocessor through a number of input ports.

Serial keyboards are self contained peripherals using a standard serial and handshaking protocol. This makes them expensive and requires the microcomputer to provide a UART or software serial interface to which it may connect, again adding to system cost.

Alternatively, a keyboard may be implemented as a matrix of switches directly connected to the microcomputer buses as a number of 8-bit input ports. This method of interfacing is simple and inexpensive, but has the disadvantage that the keyboard scanning must be performed by software, placing an additional load on the processor.

### CRT Display

CRT display interfacing is complicated and requires specialised hardware. Dedicated CRT monitors that required nothing more than a serial data feed and appropriate controlling software were available, but were expensive and offered limited graphical capability. Microcomputer manufacturers therefore often sought cheaper direct connection displays, and commonly targeted the domestic television set.

It was usual for such microcomputers to make use of dedicated display controller ICs, which interfaced to the processor as an I/O device. Typically the display controller and processor shared an area of memory known as display RAM, through which software passed information to be displayed to the display controller. Because this RAM was shared, the processor and display controller could not both access it at the same time. One device had to be given priority.

Where the processor was given priority, the display controller would be prevented from accessing the display RAM while the processor was writing to it, causing a momentary screen flicker. Such flicker could be avoided by carefully written software that wrote to the display RAM only when the display controller was not updating the display.

Where the display controller was given priority, the processor would be briefly halted while attempting to write to the display RAM if the video controller was performing a display update. This guaranteed the display would not flicker, but meant that the processor would slow down when making accesses to the display RAM. This side effect could be minimised by restricting these processor writes to periods when no display update was being carried out.

Alternatively, some microprocessors released control of memory devices at frequent and regular clock cycles, allowing microcomputers based on them to hide display controller memory accesses within these cycles.

## Architecture Evolution

This standard architecture formed the basis of the original 8-bit IBM 5150 Personal Computer of 1981, and the later 16-bit IBM 5170 PC AT that became the basis of all modern personal computers. The processor bus was supplemented by the Industry Standard Architecture (ISA) bus which abstracted I/O devices and their direct memory accesses (DMA) from the processor bus. The ROM became known as the BIOS, and contained enough program code to instruct the processor in how to load the first program from disk into RAM, and then execute it.

The architecture has changed little since then; the CPU still remains as a distinct component in its own right, the data, address and control buses have become more complicated with the advent of 16, 32 and 64-bit buses, and have been supplemented by additional system buses such as EISA and PCI. Most I/O is performed through dedicated I/O controller chips that interface directly to these buses, at clock speeds independent to that of the processor.

## The Z80 Microcomputer

The Z80 microprocessor was developed by Zilog Inc. in 1976. A more powerful processor than its predecessor, the Intel 8080, which was developed by Zilog founder Federico Faggin at Intel in 1974, the Z80 went on to become one of the most dominant processors of the late 70s and 80s.

The Z80 is binary compatible with the 8080, but provides many additional instructions and registers, as well as improved interrupt handling, 16-bit I/O addressing, simple power requirements and on-chip dynamic memory refresh control.

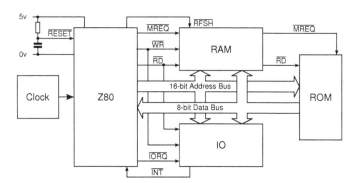

*Figure 3-2: The Z80 microcomputer*

A basic Z80 system is shown in figure Figure 3-2, illustrating the signals provided by the processor for the control of system devices.

### Dynamic RAM Interface

Dynamic Random Access Memory (DRAM) was the only type of RAM with a capacity large enough and at a cost economical enough to be used for data and program store in microcomputers of the 70s and 80s. Dynamic RAM has complicated interfacing and timing requirements due to its address bus being multiplexed, at its need to have its 'memory' refreshed every few milliseconds. See the section called *Dynamic RAM*.

One of the advantages of the Z80 over its competitors, such as the 6502, is that it provides all of the control signals, and at the necessary timing, to interface directly to dynamic RAM without the need for additional hardware[1] such as a refresh controller.

The Z80 exploits the instruction decode cycle of an instruction fetch, which does not need access to the address bus, to perform a dynamic RAM refresh. During this cycle, a 7-bit refresh address is placed on the address bus and the /RFSH control signal is activated. This tells the RAM to perform a refresh of the addressed dynamic RAM row, recharging the capacitors storing the data. The 7-bit address is incremented every instruction fetch, so that every row of the dynamic RAM is refreshed within the critical period common to dynamic memories designed at the time of the Z80.

The remaining interfacing requirement is address bus decoding, so that memory devices appear at their correct place in the memory map. For this the Z80 provides a memory request signal, /MREQ, which indicates that a memory operation is taking place and the address on the address bus should be decoded and the appropriate device selected. In addition it provides separate read and write signals, /RD and /WR, that indicate when the memory should perform the required operation.

### ROM Interface

As with dynamic RAM interfacing above, the Z80 control signal /MREQ is used to enable decoding of the address bus, and the subsequent selection of the appropriate memory device. Following this the read request signal, /RD, activates the ROM output enable, causing it to place the requested byte onto the data bus for the processor to read. A Z80 system will always find ROM mapped to the start of its address space, as the processor begins execution at address zero on powering up.

## I/O Interface

The Z80's I/O interfacing facilities mirror those for memory, and have signal and instruction timings that are tailored specifically to I/O operations.

When the Z80 wishes to access an I/O device, it places the address of the required I/O port on the address bus, and activates the I/O request signal, /IORQ. This indicates that an I/O operation is taking place and that the address on the address bus should be decoded and the appropriate I/O device selected. A short time later, one of the read and write request signals, /RD and /WR, is activated indicating that the device should perform the request.

## Interrupts

Some I/O devices are asynchronous in design, and do not require the processor to be reading or writing to them for them to be performing I/O activities. For example, a printer may be sent a line of text to print, and as printing is generally slow, the processor may get on with some other action while that is happening. For the processor to send a second line of text to the printer it needs to know when it has finished with the first. It could determine this by periodically reading an I/O port to get the current status of the printer. However, this requires repeated effort by the processor and limits the tasks it could be doing while waiting. Alternatively, the system designers could arrange for the printer to interrupt the processor when it is ready for more data, freeing the processor completely and allowing it to direct its attention to other activities.

The Z80 provides several interrupt mechanisms for use by I/O devices. A device that wishes to inform the processor that it requires attention can do so by activating the Z80 control signal /INT. On receipt of an interrupt signal, the Z80 will complete the current instruction and then begin execution of a special section of code called the Interrupt Service Routine. The ISR usually looks to see what I/O device raised the interrupt, and executes the appropriate code to handle that device. Once the interrupt has been satisfied, execution returns to the point the processor was interrupted, where it continues. The interrupted program is generally unaware that interruption took place.

## Microcomputer Implementation

Microcomputers of the 70s were typically implemented through small scale integration (SSI) using off the shelf integrated circuits such as 74LS NOR and NAND gates, counters and flip-flops. As the complexities of these computers rose, so did their size, power consumption and manufacturing costs. In order to stay competitive, computer and other original equipment manufacturers (OEM) increasingly turned to large scale integration (LSI), where more complete functional units were implemented within a single integrated circuit.

Initially the only option available to the original equipment manufacturers was through the custom design of silicon devices, an expensive and time consuming process where a positive return on investment could only be realised after large volumes of the product had been sold.

Later, semi-custom options became available which filled the gap between SSI and LSI by reducing the development cycle and cost of producing OEM specific solutions. This allowed manufacturers to increase the complexity of their products, whilst simultaneously lowering their cost, promoting competition and innovation.

---

1. A simple address bus multiplexer is the only external interfacing requirement when using dynamic RAM that provided more than 128 bytes of storage.

# Chapter 4
# Semi-Custom Devices

Prior to the advent of semi-custom devices, the OEM in need of a microelectronic solution could either use off the shelf SSI integrated circuits or invest in the development of a fully-custom LSI device.

SSI is a cost effective option for simple designs because these generic ICs are mass produced, and are therefore:

1. inexpensive to use
2. available from multiple sources, giving the OEM supplier independence
3. highly reliable

As circuit complexity rises, so does the number of SSI devices and the area of printed circuit board (PCB) required, increasing manufacturing costs. If complexity rises further, these manufacturing costs may become so great that SSI ceases to be an economic option and a custom LSI solution must be sought.

Furthermore, the OEM conscious of commercial security would be aware of how easily SSI designs can be copied, and may seek an LSI solution to protect its intellectual property.

Fully-custom design has the advantage that only the necessary functionality is implemented, removing the redundancy associated with SSI, where chip features generally exceed design requirements. Additionally, the transistor and gate structures used may be varied to optimise performance and/or the silicon area consumed.

Development of a fully-custom LSI device requires that every aspect of the silicon device be designed from the oxide masks and Photolithography process, through to the metal interconnect layers. This is a skilled and time consuming activity, which can only be reasonably undertaken by the microelectronics manufacturer. Once designed, the device can be produced on a standard integrated circuit production line.

The cost of the fully-custom device, therefore, depends largely upon the design time and the level of skilled resource required [HURSTCSIC].

Semi-custom devices neatly fill the gap between fully-custom design and SSI by reducing the expertise and time required to produce a custom design, placing it within the capability of the OEM.

The techniques used fall broadly into two categories. One, the cell-library or macrocell approach, and two, the uncommitted masterslice approach. Today, these technologies have been replaced by the structured application-specific integrated circuit (ASIC), the field-programmable gate array (FPGA) and the complex programmable logic device (CPLD).

Semi-custom devices that were designed using the cell-library approach required the generation of a full mask set, much like a fully-custom design. However, their distinguishing feature was that the custom device was constructed from a library of proven standard components. This was similar to the SSI approach, except that precisely the required number of gate, latch and flip-flop "cells" would be interconnected on the silicon to form the custom integrated circuit.

The use of a cell-library considerably reduced the design time compared will a fully-custom device, at the expense of increased chip size and possibly a less than optimum performance.

The alternative masterslice approach is divided into two categories; the uncommitted gate array (UGA) and the uncommitted component array (UCA).

A component-array consists of a slice of silicon upon which there are an array of unconnected cells, each cell containing a number of unconnected components such as transistors and resistors. These uncommitted devices were produced ahead of requirement and stored until needed, and were, thus, given the name uncommitted component array.

The alternative to the component array was the gate array, where cells operate at the functional level instead of at the component level. These functional units could be anything from simple logic gates to fully realised logic modules.

The subsequent dedication of the pre-produced arrays to a customer's specific requirements was achieved by adding a number of final metallisation layers over the array. These layers provided the required inter-cell and between-cell connections necessary to achieve the desired circuit functionality. A major advantage of the component array was that they could be configured for analogue or digital operation or both.

For the original equipment manufacturer, this method was far cheaper than producing a fully-custom or cell-library semi-custom design, as special production plant tooling was not required, and design to production times were considerably shorter. The disadvantage was that component density was usually the lowest of the three LSI technologies, and there was often a percentage of component wastage due to the limitations of connection routing via the metal layer.

## Semi-Custom Gate Technologies

Various silicon technologies have been employed by companies in the production of gate arrays, in particular bipolar transistor and metal-oxide-semiconductor (MOS) transistor.

One of the first component array manufacturers was Fairchild who in 1967 produced the worlds first two-layer metal process, 32-gate, 20 nanosecond custom DTL component-array - the Fairchild Micromatrix 4500 [MICRO-MATRIX]. This was followed by Sylvania in 1968 with a 30 cell, four gate per cell TTL array. When Motorola joined the market, they introduced 25 and 80 TTL gate arrays, with propagation delays of just 5ns.

In 1972, UK semiconductor manufacturer Ferranti Electronics Ltd introduced its bipolar Uncommitted Logic Array (ULA), an uncommitted component array. This was based on the advanced CDI process and offered an efficient cell routing capability through a single metal interconnection layer. Ferranti quickly dominated the international semi-custom device market, and by the 1980s had a 40% world market share.

Ferranti and its ULA are discussed fully in Chapter 5, *The Ferranti ULA*.

# Chapter 5
# The Ferranti ULA

In the 1960s Ferranti achieved great technological success with semiconductor products such as the Mesa transistor and the Micronor I and II DTL integrated circuits. Their commercial success was limited however, as American companies such as Fairchild, Motorola and Texas Instruments began manufacturing and marketing their devices in Britain, competing directly with companies such as Ferranti in a fledgling market.

However superior the Micronor II was at the time, companies such as ICT (which later became ICL) purchased the cheaper Texas Instruments 5400 and 7400 TTL ICs for their 1906 computer series, despite the fact they had propagation delays of 10ns, compared to Micronor II's 8ns.

In order to reach this level of technical achievement, Ferranti were naturally reliant upon patent licences from Bell Labs, Texas Instruments and Fairchild as the early innovators of silicon technologies. Much of the funding for this and Ferranti's research and development activities came from government loan schemes such as the Advanced Computer Techniques Project, launched by the UK Ministry of Technology in 1966.

It was this availability of funds that allowed Ferranti to pursue and develop the CDI process, paying Fairchild £150,000 for the use of its patents [WILSONTT]. Though invented by Bell Labs in 1969 [MURPHY], both Bell Labs and Fairchild had failed to turn it into a viable semiconductor technology, managing only to produce prototype devices that operated at three volts, and not at the industry standard of five volts.

Towards the end of 1969 Ferranti engineers Alan Bardsley and Dick Walker visited Bell Labs to learn all the could about the CDI process and technical issues that had been experienced. On returning to the UK they reworked the entire process, and by early 1970 they had improved production yields at each stage of manufacture and increased the operating voltage to five volts. At this point of development Ferranti began planning a range of CDI integrated circuits and applications, setting out a strategy based on this, their latest, and highly significant technical achievement.

In June 1970 the Labour government lost the general election to the Conservative Party, who were keen to reduce the interference of government in industry. In the months that followed, they abolished the ACTP scheme and effectively ended the availability of credit on which Ferranti and others had been so reliant [CPMAN1970][WILSON2]. Coupled with diminishing sales on standard ICs, Ferranti was clear that it would be unable to compete in price with their U.S. competitors, and was forced to review its product portfolio and strategies in a drive to remain in the semiconductor manufacturing arena.

Inspired by their technical strengths, Ferranti adopted a strategy of finding a niche market for their new advanced CDI process. One such niche was the manufacture of non-standard ICs for original equipment manufacturers who required custom designs.

It was this direction of research that gave birth to the Uncommitted Logic Array (ULA). This unique bipolar semi-custom LSI device could be customised to an individual customer's requirements quickly and cheaply, offering the most cost effective technique for designing and producing LSI for customer specific applications.

In 1972 Ferranti Semiconductors pioneered and introduced the first commercially available ULA, providing an economic large scale integration solution to a customer's requirements in a fixed, fast and dependable timescale, whatever the market sector [FERRANTISG1].

The following year Ferranti received an unexpected boost in the form of renewed funding from the UK Department of Trade and Industry and its Microelectronics Support Scheme. This allowed Ferranti to claim 50 per cent of all research and development costs and enabled it to perfect the CDI process even further, significantly improving production reliability.

ULA's were initially produced in two categories, a digital/linear optimised ULA Digilin Array using RTL technology, and a ULA Gate Array using CML with increased operating speeds and packing densities. Like other component arrays, each ULA contained an array of 'uncommitted' active and passive components, but uniquely required only a single final layer of interconnecting aluminium which allowed a high degree of chip utilisation and complexity.

The interconnection pattern connecting the individual components to provide system integration was generated from the customer's own specification either by Ferranti or by the customer themselves.

By the end of the 1970s, the ULA family was composed of 12 arrays offering complexities from 200 to 1000 gates, with speeds of up to 20MHz. These were marketed as the ULA1000, 2000 and 5000 series, with each providing a range of performance and power ratios coupled with flexible digital and analogue capability.

By 1982 the family had increased to 50 arrays, following the introduction of a new buffered-CML technology. These new 'R' series arrays offered up to 10,000 gates at frequencies up to 80MHz, with low power consumption in four performance and power ratios.

Ferranti also began offering the automated ULA CAD Complex services at this time, based on the Applicon AGS860/870 interactive graphics and the VAX11/780 computer system supported by PDP11/60, PDP11/05, design stations and high speed plotters. This facility automated the routing of ULA metallisation layers to VLSI complexity, and was linked to the automatic ULA test centre and microlithography centre.

Over 40 man-years of dedicated gate array software was available, with an extensive design library and all the programs necessary for ULA design and verification, including layout and design rule checking, logic validation and simulation, high level test language, test program generation and test schedule verification. The ULA Designer was the first system of its kind, worldwide, that could be installed at a customers premises, allowing an engineer to control and carry out the full ULA design cycle up to the manufacturing stage. This powerful multi-user system consisted of a DEC PDP11/23 minicomputer running the RSX-11M operating system, high quality graphical display, digitising tablet, printer/plotter and control console. A verified design for a 1000-gate array would typically take three to four weeks, with prototypes available after an additional four to seven weeks. Verification was achieved by communicating with the Ferranti CAD Complex using a 2400 baud dial-up modem using the DECNET protocol. The ULA Designer was announced on the 23rd February 1982, at a base price of $99,000.

## ULA Organisation

Each ULA chip has a matrix of identical cells containing uncommitted components, whose role is to satisfy the functional logic of the system, surrounded by peripheral cells of uncommitted components for input/output and linear functions.

**Matrix Cells**

The number of matrix cells on any given array determines the logic complexity that can be achieved.

Each matrix cell contains a number of unconnected components, which when connected in their basic form provide two 2-input NOR gates.

There are three types of matrix cell, RTL, CML and buffered CML, each of which provide performance advantages depending on the particular application.

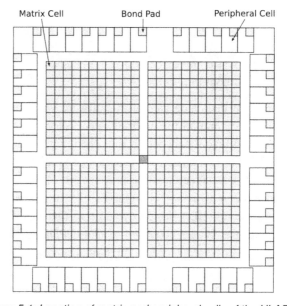

*Figure 5-1: Location of matrix and peripheral cells of the ULA5000*

**Peripheral Cells**

Arranged around the periphery of the chip, the peripheral cells provide I/O and linear functionality. The number of peripheral cells available is governed by the physical chip dimensions, which is related to the number of matrix cells.

ULA Gate Arrays have peripheral cells whose components provide I/O interface capability with bipolar, MOS and CMOS technologies. They may be configured to drive LCD and LED displays and switch outputs, and in addition will provide linear functions such as oscillators and Schmitt triggers.

ULA Digilin Arrays (those with combined digital and enhanced linear capability) have peripheral cells whose components provide high performance linear functions such as amplifiers, precision analogue switches, comparators and op-amps, as well as digital I/O. In addition, the later generation of Digilin Array peripheral areas contain special supporting linear elements such as voltage reference sources, high current drive transistors and shaping capacitors.

## ULA Design Process

Prior to the availability of the ULACAD system, the ULA design process was relatively manual, though simple and efficient. The later CAD process followed the same key steps, but through automated tools, reducing the turn around times and producing more automatically correct designs.

The general sequence is as follows:

1. Determine the type of array, define and agree the logic specification.
2. Generate a single interconnection pattern from agreed specification.
3. Produce single interconnection mask.
4. Fabricate prototype samples by applying aluminium interconnection patterns to uncommitted wafers held in stock.
5. Prototype testing.

Steps one to three could be carried out by the Ferranti ULA engineering team or by the customer. Fabrication and production testing was carried out by Ferranti, and usually followed by a level of customer testing.

### Determine Array Type and Agree Logic

The choice of ULA depends on the complexity of the logic to be implemented, once it had been designed down to the gate level. Deciding factors would be the amount and accuracy of the linear function required, the number of pins and cells needed, the power requirements and propagation delays.

Since each ULA array provided different internal component values, the choice of ULA array and final circuit design were closely related. It was usual for a suitable array candidate to be chosen before the design was finalised, so it could be expressed in terms of the component values and propagation delays found in the selected array.

Once a suitable array type had been chosen, the logic and linear circuit design would be prototyped on breadboard or simulated. Once verified, the design would be agreed with Ferranti and signed-off. This was particularly important if Ferranti were carrying out the design and development life cycle.

### Generate Interconnect Pattern

Converting the agreed logic diagram into an interconnect pattern was the most complex and time consuming task. A manual $250\times$ magnification drawing of the interconnect pattern was created in pencil on mylar, laid over a routing grid identifying the matrix and peripheral cell component connection points. The approach used to place the logic gates on the matrix was to divide the logic diagram into blocks of strong affinity. These blocks would then be placed on the array matrix, leaving room for inter-block routing. Detailed gate placement within the blocks would then be performed.

*Figure 5-2: Generation of 250× ULA interconnection pattern*

The interconnect co-ordinates of the $250\times$ interconnect pattern was digitised into Ferranti's Applicon system and verified before being used to produce magnetic steering tapes for the final stage mask generation. The mask artwork

would then be automatically cut and then checked, producing a negative 250×
light-field mask, where significant parts of the mask were opaque. The mask
was then inspected for flaws (Figure 5-3) before being optically reduced to a
10× magnification mask called a reticle (see Figure 5-4).

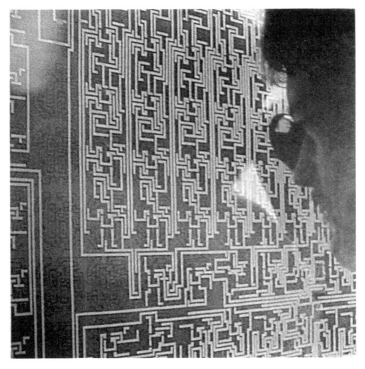

*Figure 5-3: Inspection of machine cut 250× artwork*

The main problem with a user supplied pencil interconnect pattern occurred at
the verification phase, as the user had to allocate time to visit Ferranti to solve
any problems found by the system.

Finally the single reticle was used to create a 1× magnification multi-chip
wafer mask of 552 individual masks, using an optical step-and-repeat procedure, through optical projection and a precision stepper table.

**Prototype Fabrication**

The final full-size wafer mask (Figure 5-5) containing the grid of 23×24 individual masks was used to optically expose an entire wafer.

The uncommitted wafer was first coated with a layer of aluminium and a UV sensitive negative photoresist. A few microns above this was placed the full-size wafer mask, which was then exposed to UV light. The chemical bonds in the negative photoresist strengthen under UV so that the areas not illuminated can be etched away, leaving the aluminium interconnection tracks.

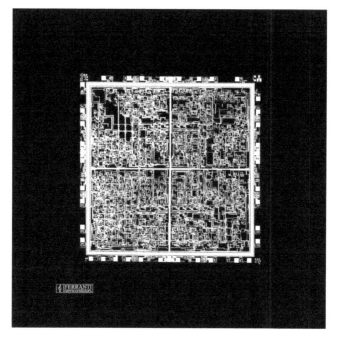

*Figure 5-4: Final 10x optical interconnect mask or reticle*

**Testing**

Before the prototype committed arrays were assembled and packaged, some verification of the individual integrated circuits would be performed. This was followed by functional testing by Ferranti and then the customer. The essential checks being [HURSTVLSI]:

- *Fabrication Checks*: Test that all the fabrication steps have been implemented during the wafer manufacture.

- *Design Checks*: Test that the prototype ICs are functionally correct and meet the specification.

• *Production Checks*: Test that production ICs to be used in the final product are free from defects.

Fabrication checks were carried out by Ferranti through special drop-ins distributed across the wafer. Drop-ins, or process evaluation devices, are small circuits that allow various parameters of a wafers fabrication to be checked such as resistivity, transistor junction breakdown voltage and capacitance. These features were checked before any functional verification of the surrounding committed array dies was carried out. Five drop-ins can be seen arranged in a cross pattern over the full-size interconnect mask in Figure 5-5.

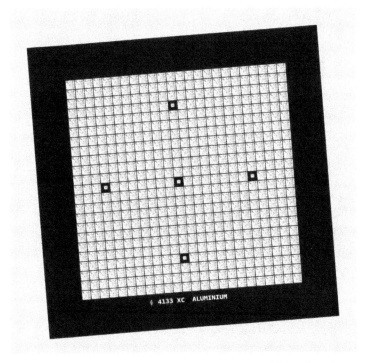

*Figure 5-5: Full-size wafer interconnection mask with five drop-ins*

Customer testing of prototypes was essential, as there was no guarantee that a committed array that passed Ferranti's production and functional tests would work correctly when subjected to real product conditions by the customer. These may not be production faults, but minor design or specification errors that occurred due to changes in the product specification after the ULA logic design had been agreed, or from incomplete design verification and simulation at the initial design phase.

Production testing of committed arrays was always carried out, as the fabrication of integrated circuits can never be considered defect free. Each die on the wafer was individually tested through an automated wafer probe that rapidly applied the customer supplied test vectors. Those dies that failed were painted with an ink spot so that they could be discarded after dicing.

## ULA Computer Aided Design

From approximately 1981, Ferranti offered computer aided design and verification of gate array designs at their ULACAD complex [RAMSAYAUTO], and from 1982 at the customers own premises through the ULA Designer system.

Both systems implemented the same design sequence consisting of four simple steps, which were straightforward enough that no expertise was required to use the ULA Designer [FERRANTIUDB]:

1. *Enter Logic Diagram*: The logic diagram, the key reference database for layout and logic verification routines, was digitised into the ULA Designer as a complete graphical drawing and syntax checked.

2. *Design ULA Layout*: The user carried out the layout design using standard and user-defined functions. When complete, the layout data was digitised into the ULA Designer and a check plot produced. Interactive editing of any errors or modifications could be performed at the graphics terminal where special ULA window and backcloth facilities were available to help the operator determine the locations of contact holes on the matrix.

3. *Enter Test Schedule*: The test schedule was written in Ferranti's high-level SAM Integrated Circuit Testing Language, which was used by the ULACAD Applicon host computer. After it had been entered into the ULA Designer and a complete syntax check performed, the schedule would be used for both logic verification and for production of the final ATE program [RAMSAYREM].

4. *Verify Design*: For this stage the user would transmit three files via modem to the ULACAD centre for processing: logic, layout and test schedule, along with processing commands (ULASIM for logic simulation, ULACHECK for layout checks and ULATEST for test program generation). Ferranti would automatically process each file with the appropriate program and return the results to the users ULA Designer

later that day. The user therefore had full access to the specialised ULA software necessary for complete design verification.

Once the design was complete, the user would transmit instructions to the ULACAD complex to proceed with mask and prototype device manufacture.

Logic diagram entry was considered to be one of the key stages of the automation process, as nearly all subsequent stages of the process, from logic simulation and interconnect routing, through to layout checking, would be performed by referring to the logic design and related information.

*Figure 5-6: Digitising logic diagram into the ULA Designer*

The logic diagram produced by the CAD system served as the master reference database, once it had been checked by the original designer. Several items of reference data were required to be held against the logic drawing:

1. Network interconnections, generated through the digitisation process.

2. Gate types.

3. Gate propagation delays.

4. Gate supply currents.

43

5. ULA type.

6. Peripheral cell type (for instance the ULA2U000 array contains two different types of peripheral cell).

7. Peripheral cell propagation delays.

8. Peripheral cell supply currents.

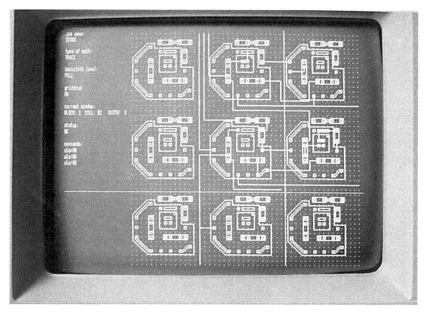

*Figure 5-7: Interactive editing of layout in the ULA Designer*

Once layout paths had been stored in the ULA Designer or ULACAD system, they could be edited or modified through the interactive graphical terminal, Figure 5-7. Such modification could be carried out during the digitising process, allowing immediate correction.

The design verification stage automated the process of layout checking by comparing against the logic drawing master database. Since this was performed at the ULACAD Complex, layouts produced using a remote ULA Designer were required to be transmitted to Ferranti by modem link.

Based on experience gained through the 1970s, Ferranti found that most routing errors were between components within a cell. Ferranti therefore split the layout checking into two phases:

1. *Intra-cell checking*: Verify that each logic cell is wired correctly in terms of transistors and resistors. This was necessary due to the possibility of errors in manual or hand finished routing.

2. *Inter-cell checking*: Verify that logic cells are correctly routed against the logic diagram.

Designs that were carried out by Ferranti engineers at the ULACAD complex could be laid out using automatic logic placement and routing. Automatic placement of a layout followed the same process as manual placement, in that the logic was split into a few blocks of strong affinity which were placed allowing for inter-block routing. Once the blocks had been placed, cell orientation had to be considered. A ULA cell may be routed in multiple ways to perform the same logic function, the complexity of which increased as you went above a 4 input NOR gate, such as a D-type latch which could require between four and six matrix cells. The CAD system employed a *center of gravity tree* calculation for each net to calculate the best configuration.

Automatic routing of the array was carried out using algorithms derived from the Applicon PCB routing application. The common conduction routes of the matrix cells, such as cross-unders and the path between multiple common collectors of a transistor, were held as fixed routes on a virtual second routing layer, so that the router had the necessary information about pre-connected points.

One hundred per cent full automatic routing was generally not achieved due to the high packing density required on most ULA array designs. Some degree of hand finishing was therefore required on many layouts, often involving modification of internal cell routing to achieve a fully interconnected design.

## ULA Construction

The ULAs discussed in this chapter were produced using the Ferranti Advanced Bipolar LSI process (FAB-2), a successful implementation of the Bell Labs/Fairchild CDI process discussed in Chapter 2, *Integrated Circuits*.

This semiconductor process utilised N+ diffusions to provide both isolation as well as transistor base and resistor areas. By employing a thin epitaxial layer of $2\mu$m and shallow diffusion depths of $1\mu$m, a component packing density equivalent to NMOS was achieved [ULAHAND].

Figure 5-8 shows an exploded view of the FAB-2 construction of a ULA transistor and resistor.

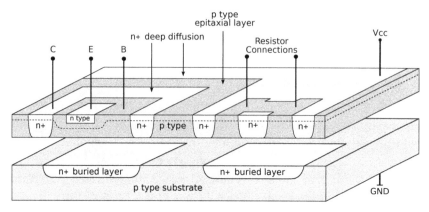

*Figure 5-8: Exploded FAB-2 CDI transistor and resistor*

The entire p-type substrate is used to provide the ground rail for the ULA, being brought to the die surface by the p-type epitaxial layer which surrounds transistor areas. Vcc on the other hand is provided by highly doped n+ diffusions, such as that surrounding resistors, allowing the ULA supply connections to be made without metal, and the single aluminium layer to be used entirely for component interconnection. As ULA sizes increase, Vcc routing features are incorporated into the array to ensure an even distribution of supply voltage across the ULA.

### Matrix Cells

There were three types of ULA matrix cell in use by 1982:

1. *RTL* or Resistor-Transisor Logic cells
2. *CML* or Current-Mode Logic cells
3. *BCML* or Buffered Current-Mode Logic cells

Resistor-Transistor Logic cells were used by the first available ULA arrays. The resistor values used in each type were carefully chosen to provide a particular speed to power ratio. Diffused resistors were generally used, except for values over 10K which were implemented as pinch resistors. Figure 5-9 shows the components of an RTL matrix cells, typically with Rin and RL of 10K for standard array types, and of 120K for low power array types.

A wide supply voltage was supported by RTL arrays, and the matrix cells themselves could operate between 1.0v and 5.0v.

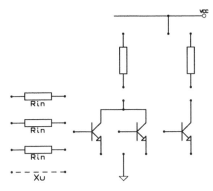

*Figure 5-9: ULA 2000 RTL matrix cell schematic*

Current-Mode Logic offered a faster switching speed and lower current consumption than RTL, and was used in the later ULA gate arrays as the range was extended. Like RTL cells, the resistor values of each CML array type configured a particular speed to power ratio. Figure 5-11 shows the ratio operating frequency to gate current for the RTL and CML arrays of the ULA 1000–5000 series. CML arrays supported a wide supply voltage but as their matrix cells operated at a fixed voltage of between 0.84 and 0.95v, current consumption was considerably reduced.

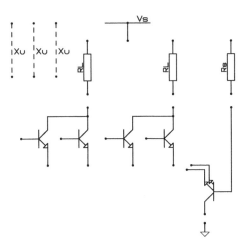

*Figure 5-10: ULA 2C000 CML matrix cell schematic*

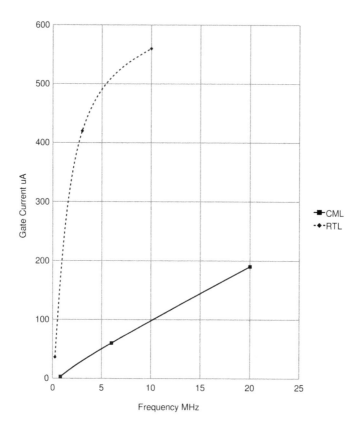

*Figure 5-11: RTL and CML frequency to current ratio*

The operation of CML circuits was discussed in Chapter 2, *Integrated Circuits*. The simplest configuration of the CML cell is that of a NOR gate provided by the common-collector switching transistors. The current source and load resistors defining the voltage swing between logic states, and also the power consumption of the gate. Low power device types therefore employed high value resistors in the matrix cells; 90K in the case of the 2U000 type.

Figure 5-12 shows the voltage levels within half a 5C000 matrix cell, configured as a two input NOR gate.

Buffered current-mode logic cells were first used at the end of 1981 with the introduction of the R series ULA. These matrix cells contained CML gates with buffered outputs that increased the fan-out of each gate, and resistor values that when combined with an improved manufacturing process, improved the switching speed to power product ratio by a factor of 4 to 1 (Figure 5-13).

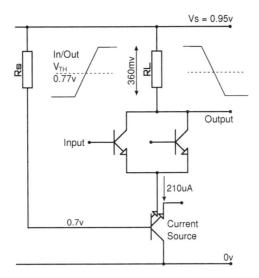

Figure 5-12: CML NOR gate logic thresholds

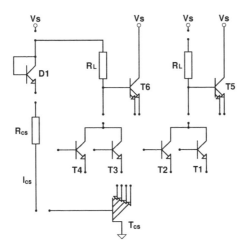

Figure 5-13: R Series buffered CML matrix cell schematic

## Peripheral Cells

External interfacing and linear functionality is provided by specially designed peripheral cells arranged around the periphery of the array. The complexity of these cells and the component values used was determined by the intended use of the array, and were specifically matched to the matrix cells.

There were three general configurations of peripheral cell depending on the type of array: one for RTL arrays, another for CML arrays and a third for DIG-ILIN arrays. The component values used by an array established its switching speed and power consumption and therefore differed between ULA array types of the same family.

RTL peripheral cells were of limited complexity and generally had high value pinch resistors.

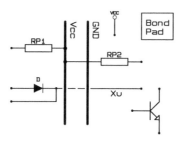

*Figure 5-14: ULA2000 RTL peripheral cell schematic*

CML peripheral cells had a high complexity and a good spread of low value resistors. Figure 5-15 shows a 2C000/5C000 series peripheral cell.

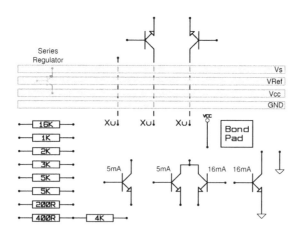

*Figure 5-15: ULA2C000 CML peripheral cell schematic*

The Digilin peripheral cells contained high performance components such as matched transistors and a good spread of high value resistors, contributing to their low power consumption. For instance, the 2U000 very low power CML device contained 40K, 80K, 160K and 500K resistors. In addition, the 2U000 device provided two different complexities of peripheral cell, avoiding component wastage where simpler I/O was required.

**Power Rails**

In a fully utilised array the largest power consumers would be the peripheral cells. Because of this, the power rails were routed along the inside edge of the peripheral cells, surrounding the central matrix square, as shown in Figure 5-16. Not only did this provide power precisely where it was needed, but was the most efficient way to supply the matrix cells by guaranteeing that no cell was further than half the matrix width from the Vs supply. These main power rails were the only fixed metal interconnections required in a ULA design, as shown schematically in Figure 5-15.

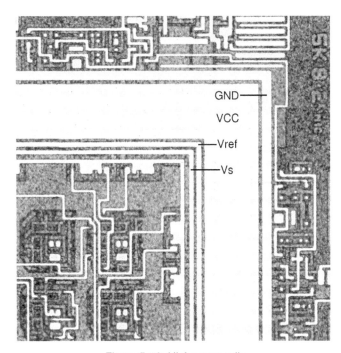

*Figure 5-16: ULA power rails*

Peripheral cells operated from Vcc whereas matrix cells operated from Vs which, in the case of CML arrays, was a regulated noise-free supply of between 0.84v and 0.95v generated by temperature compensated series-regulators located at the base of each peripheral cell. These regulators drove the Vs rail from Vreg (usually Vcc), and were controlled by a Bandgap reference voltage of between 1.35 and 1.5v provided by discrete components at two opposing corners of the chip. Since there are many Vs regulators evenly distributed around the outside of the array, the total current available was much greater than a single regulator could provide. Furthermore, as the number of cells in the matrix increased, so did the circumference of the array. This allowed more peripheral cells and series regulators to be fabricated, which in turn provided the additional current required by the increase in matrix cells.

All arrays could be powered from a single supply, where the internal Vcc and regulator supply Vreg were commoned. Alternatively, CML arrays permitted the use of two separate supply voltages for Vcc and Vreg, producing a lower power dissipation without compromising speed.

The basic chip organisation of a CML ULA is shown in Figure 5-17.

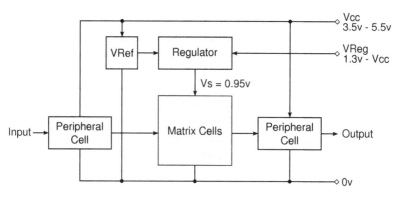

*Figure 5-17: Basic CML chip organisation*

## The ULA1000, 2000 and 5000 Series

The ULA1000–5000 series of arrays were developed from the first ULA introduced in late 1972, and by 1979 offered complexities from 200 to 1000 gates at a range of power and performance levels (see Figure 5-11). Early low-speed versions were implemented using RTL and were later supplemented by faster, lower power CML types.

The speed and power ratings that were available are shown in Table 5-1, the type code determines the ULA device. For instance a 5N000 ULA would be a Low Power, Medium speed 5000 series ULA.

| Type Code | Description | Clock (MHz) | Gate Delay (ns) | Gate Current ($\mu$A) | Logic Type |
|---|---|---|---|---|---|
| L | Low Power Digilin | 0.250 | 200 | 36 | RTL |
| U | Very Low Power Digilin | 0.800 | 450 | 3 | CML |
| - | Normal | 3 | 25 | 420 | RTL$_1$ |
| N | Low Power Medium Speed | 6 | 25 | 60 | CML |
| H | High Speed | 10 | 10 | 560 | RTL |
| C | Very High Speed | 20 | 8 | 210 | CML |

*Table 5-1: Ferranti ULA1000–5000 speed and power ratings*

Each array in the series has its own geometry, accommodating the varying complexities available. The ULA1000 and ULA2000 arrays both contain a single central matrix, however the much larger ULA5000 matrix may be divided by the Vs supply rail into four quarters, shown in Figure 5-18.

| Series | Types | Matrix Size | Matrix Cells | Peripheral Cells | Average Gates |
|---|---|---|---|---|---|
| 1000 | - L H | $10 \times 10 \times 1$ | 100 | 28 | 150 |
| 1000 | U | $11 \times 13 \times 1$ | 143 | 26 | 215 |
| 2000 | - L H N C | $15 \times 15 \times 1$ | 225 | 40 | 337 |
| 2000 | U | $16 \times 16 \times 1$ | 256 | 40 | 384 |
| 5000 | - ($_1$) | $11 \times 11 \times 4$ | 484 | 48 | 726 |
| 5000 | L | $22 \times 22 \times 1$ | 484 | 48 | 726 |
| 5000 | N C | $11 \times 10 \times 4$ | 440 | 48 | 660 |

*Table 5-2: Ferranti ULA1000–5000 array sizes*

The size and cell count for each array type is given by Table 5-2.

The simple coupled transistors and resistors in a single ULA cell were not very useful on their own other than for implementing a basic switch. To create the more common logic functionality used as the basic building blocks of more complex designs, ULA cells must be combined and interconnected. The more complex the logic function, the more ULA cells required. For instance, NOR and NAND gates are the simplest 'complete' gates and require single cells, flip-flops and counters on the other hand require far more. This is illustrated by Table 5-3 for a 5C000 ULA device.

| Logic Function | Cell Count |
|---|---|
| TTL Buffer | 1 |
| 3 Input NOR Gate | 1 |
| 2 Input NAND Gate | 1 |
| Monostable | 2 |
| Binary Divider with Preset and Clear | 3 |
| Data Latch | 3 |
| D-Type Flip Flop with Preset and Clear | 6 |
| Differential Amplifier | 6 |

*Table 5-3: Cell count for typical logic functions*

**The 5000 Series ULA**

The 5000 Series ULA is show in Figure 5-18. The matrix cells are arranged as four blocks of $11 \times 11$ cells[1], separated by a channel containing six cross-unders per matrix row and column. Running over this channel, above the cross-unders, is the Vs power rail which distributes power evenly across the matrix. The exception is the 5L000 device which contains a single block of 22 $\times$ 22 matrix cells, being a low power type it does not require elaborate supply routing, and better use is made of available space by have additional matrix cells rather than power routing channels.

Peripheral Cell          Matrix Cell          Bond Pad

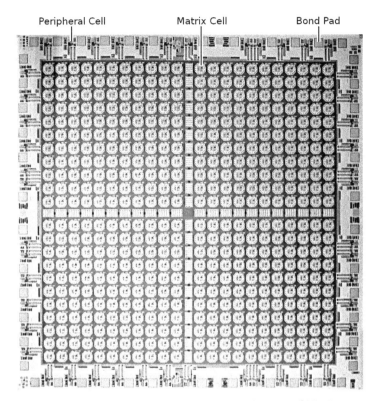

*Figure 5-18: Uncommitted ULA 5000 standard type CML die*

The ULA 2000/5000 CML matrix cell has the structure shown in Figure 5-19, which is described by schematic Figure 5-20 and contains the following components:

- Two pairs of collector coupled transistors
- A dual current source supplying 120 $\mu$A (T5)
- Two load resistors (RL): ULA Type N = 5K7, Type C = 1K7
- One current source resistor (Rcs): ULA Type N = 5K7, Type C = 1K7
- A Vs supply connection
- A GND connection
- Three cross-unders

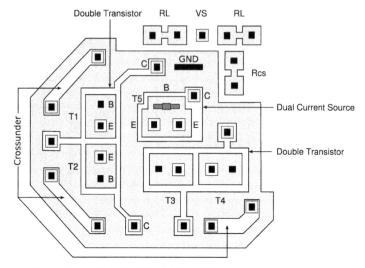

*Figure 5-19: ULA 5C000 / 2C000 CML matrix cell*

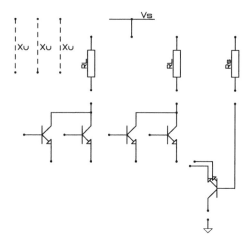

*Figure 5-20: ULA 5C000 / 2C000 CML matrix schematic*

Surrounding the central gate-array cell matrix are the peripheral cells, arranged around the edge of the chip. These cells are similar to the matrix cells but have a higher current handling capacity and a larger component count, making them suitable for linear and interfacing functions. Each peripheral cell contains a bond pad for connection to the external pin of the IC package. There may be between 26 and 48 peripheral cells, depending on chip size, but up to a maximum of 40 package pins, determined by the choice of DIL package.

The ULA 2000/5000 CML peripheral cell contains:

- Two coupled transistors rated at 5mA and 16mA
- One transistor rated at 16mA
- One transistor rated at 5mA
- Two general purpose transistors
- A 16K pull-up resistor
- Six resistors of value 200R, 1K, 2K, 3K, 5K and 5K
- One voltage divider of 4K and 400R
- A Vs supply rail at 0.84 or 0.95v, depending on ULA type
- A Vcc supply rail at between 3.5 and 5.5v
- A reference voltage rail
- A ground rail
- Three cross-unders
- An IC pin bond pad

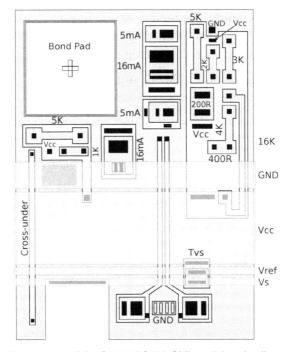

*Figure 5-21: ULA 2C000 / 5C000 CML peripheral cell*

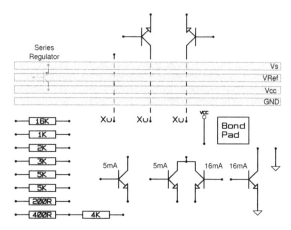

*Figure 5-22: ULA 2C000 / 5C000 CML peripheral cell schematic*

Note that the Vs rail at the base of each peripheral cell connects directly to the n+ diffusion that surrounds all matrix and peripheral cells, providing an evenly distributed supply (white area at bottom of Figure 5-21). Figure 5-19 shows the matrix cell Vs connection attached to the same layer. The GND rail is similarly distributed by bonding to the p-type epitaxial layer at each peripheral cell (grey), where it is carried through to the substrate.

The power dissipation of a 5C000 ULA is given by:

$$ULADissipation = MatrixDissipation + PeripheralDissipation$$

where the matrix cell dissipation is 0.95mW per gate, and the peripheral cell 11mW for a Vcc of 5v.

A 5C000 ULA powered at 5v utilising 300 active gates and 40 peripheral cells has a total power dissipation of:

$$(300 \times 0.95mW) + (40 \times 11mW) = 725mW$$

## The R Series ULA

Developed from the range of digital CML arrays ULA2N000 through to ULA9C000, the R Series ULAs improved the speed to power product by a factor of 4 to 1, reducing the current consumption of a gate, increasing switching speed, improving peripheral cell capability and reducing the number of matrix cells required to implement a design [FERRANTIRS].

They were initially available in three speed/power options: high speed type RA, offering 2.5ns typical gate delay, standard type RB offering 7.5ns gate delay, and a low power type RC with $30\mu W$ per gate, 15.0ns delay (Figure 5-23). Each came in five configurations offering between 500 and 2000 gates. Later, the lower speed and power RD type was added, offering $8\mu W$ per gate.

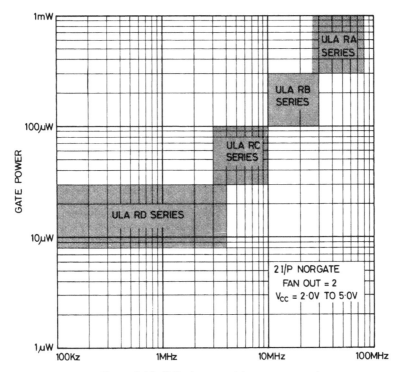

*Figure 5-23: R Series speed / power comparison*

The matrix cells were more advanced that the earlier CML series, having an additional emitter follower added to each gate with two electrically isolated emitter outputs. The dual gate outputs permitted a wired-OR arrangement which reduced the gate count required by the earlier ULA types.

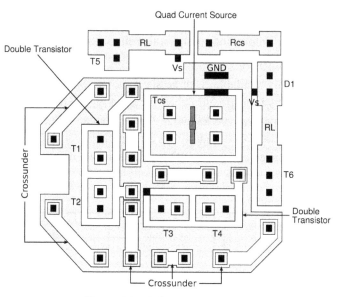

*Figure 5-24: R Series matrix cell*

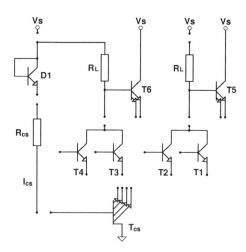

*Figure 5-25: R Series matrix cell schematic*

In total the R Series matrix cells contain six transistors, one diode, a quad current source and seven cross-unders. A typical dual 2-input NOR gate configuration, supporting two independent outputs per gate is shown in Figure 5-26.

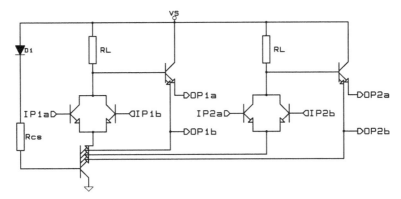

*Figure 5-26: R Series matrix cell configured as a dual 2-input NOR gate*

The peripheral cells were also more advanced as they contained 11 resistors and eight transistors. The resistor values themselves differed between R series types, providing the range of power and speed options. With lower values the ULA switched faster, but at the expense of power consumption. The resistor values for each type are shown in Table 5-4.

The R Series peripheral cell shown in Figure 5-27 contains the following components:

- Three cross-unders
- Three Vcc connections
- Three GND connections
- An IC pin bond pad
- A Vs supply rail at 0.95v
- A Vcc supply rail at between 3.5 and 5.5v
- A reference voltage rail
- A ground rail
- A series regulator to generate the matrix cell supply Vs
- Four transistors T1, T2, T3, T6 rated at 7mA, 6mA, 6mA and 16mA
- Two collector coupled transistors T4 and T5, rated at 6mA and 18mA
- Two general purpose transistors T7 and T8, rated at 11mA
- One voltage divider of 500R and 1K5
- Nine resistors, Rp1-9, the values of which depend on ULA type

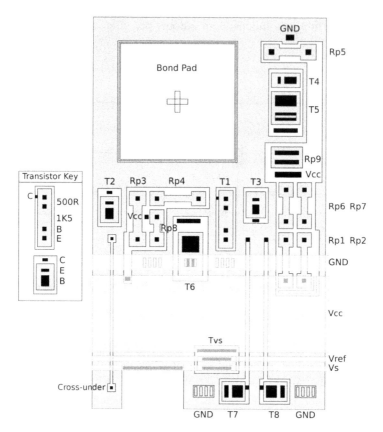

*Figure 5-27: R Series peripheral cell*

| Resistor | RA | RB | RC |
|----------|------|------|------|
| Rcs | 3K | 11K5 | 30K |
| RL | 2K6 | 10K | 26K |
| Rp1–4 | 4K1 | 7K2 | 45K |
| Rp5 | 3K6 | 6K3 | 38K |
| Rp6,7 | 3K1 | 5K4 | 32K |
| Rp8 | 2K1 | 3K7 | 20K |
| Rp9 | 100R | 100R | 100R |

*Table 5-4: R Series matrix and peripheral cell resistor values*

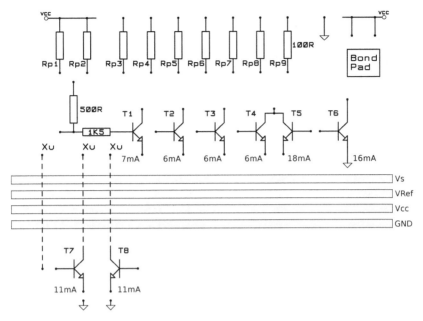

*Figure 5-28: ULA R series peripheral cell schematic*

## The 6000 Series ULA

The 6000 series ULA advanced the 5000 series by incorporating features from the R series ULA, and appears to have been produced specifically for Sinclair Research Limited to reduce the ZX Spectrum's power consumption and heat output.

Ferranti product selection guides from the early to mid 1980s, when Sinclair was using the 6C001 ULA, provide several array comparison tables between early ULA2C and later ULA12C devices and the R-series arrays. At no point is reference made to the ULA6C array (Figure 5-29).

Sinclair was Ferranti's single largest user of ULAs [WILSON2] and it is conceivable that a special array satisfying Sinclair's requirements could have been produced from Ferranti's existing technologies, considering the high production volumes that Sinclair would have required.

The 6C000 matrix cell is almost identical to the 5C000 in structure and layout. This would have been important to reduce the amount of redesign and re-routing required. Physically the dual-current source has been rotated 90° clockwise and the resistors grid aligned to the transistor connections. Additionally, Rcs and RL2 share a common connection, adjacent to which a second Vs connection is provided. The overall structure is shown in Figure 5-30.

## System Speeds to 20MHz

| Array Type | Gate Count | Gate Delay | Gate Power | IO Cells | Max Pin |
|------------|------------|------------|------------|----------|---------|
| ULA2C | 450 | 8.0 | 250 | 40 | 40 |
| ULA5RB | 500 | 7.5 | 100 | 38 | 40 |
| ULA5C | 900 | 8.0 | 250 | 48 | 52 |
| ULA9RB | 900 | 7.5 | 100 | 48 | 50 |
| ULA12RB | 1200 | 7.5 | 100 | 52 | 58 |
| ULA16RB | 1600 | 7.5 | 100 | 62 | 68 |
| ULA18RB | 1800 | 7.5 | 100 | 64 | 72 |
| ULA9C | 2000 | 8.0 | 120 | 64 | 64 |
| ULA20RB | 2000 | 7.5 | 100 | 72 | 80 |
| ULA24RB | 2400 | 7.5 | 100 | 80 | 88 |
| ULA12C | 2400 | 8.0 | 120 | 68 | 70 |
| ULA30RB | 3000 | 7.5 | 100 | 82 | 96 |
| ULA40RB | 4000 | 7.5 | 100 | 118 | 130 |
| *ULA100RB | 10000 | 7.5 | 100 | 138 | 150 |

*Figure 5-29: Ferranti Product Selection Guide comparison chart*

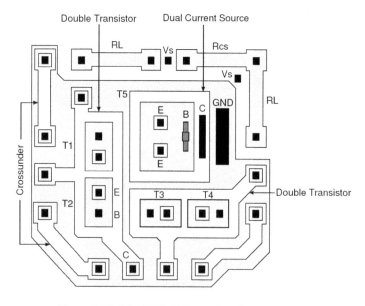

*Figure 5-30: ULA6000 CML matrix cell*

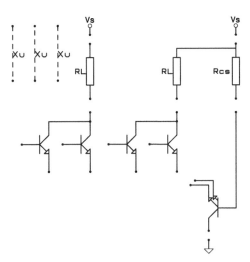

*Figure 5-31: ULA6000 matrix cell schematic*

The 6C001 peripheral cell is identical to the R Series ULA peripheral cell, except for a slight repositioning of the cross-under (Figure 5-32). It therefore has the schematic shown in Figure 5-28. Which R Series type it is based on can be determined by examining the values of resister used.

As there is no documentation available for this ULA, some elementary reverse-engineering is required to discover the resistor values used.

Ferranti's original CDI process used for the 2C and 5C devices incorporates a p+ skin diffusion that provides a sheet resistance of 470 ohms per square (Figure 2-10); therefore for a given length and width of a resistor formed from such an isolated p-type region, the effective resistance may be calculated as:

$$R = 470 \times (length \div width)$$

The 6000 series ULA may have a different sheet resistance if it was constructed from the later FAB2 process of the R Series ULA. As all R Series peripheral cells have a 100 ohm resistor in common, by measuring the dimensions of the equivalent resistor in the 6C001 ULA, Figure 5-33, the effective sheet resistance can be calculated. This allows each resistor value of the 6000 series ULA to be derived without knowing the R Series type on which it is based.

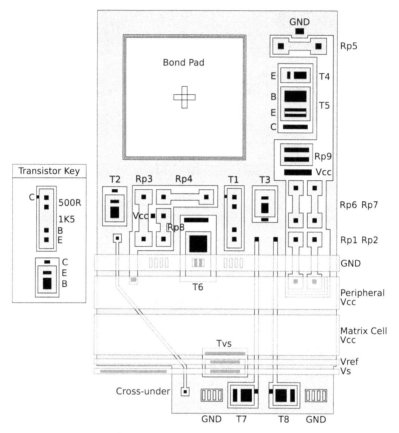

*Figure 5-32: 6C001 peripheral cell*

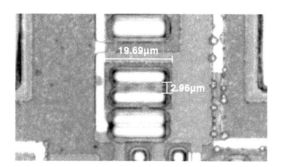

*Figure 5-33: 6C001 peripheral cell 100R resistor dimensions*

Rearranging the previous equation in terms of resistance and dimension, and substituting the width and height of the 6C001 ULA 100R resistor, Rp9, gives an approximate sheet resistance of:

$$R/Sqr = (100 \times 19.69\mu m) \div 2.96\mu m = 665/Sqr$$

Analysing the dimensions of the other resistors in the peripheral cell gives Rp1-4 at 4189 ohms, Rp5 at 3613 ohms, Rp6 and Rp7 at 3103 ohms and Rp8 at 2105 ohms. These values match the RA ULA type exactly. This comes as no surprise considering the RA peripheral cell resistor values are very close to those of the 5C000 ULA, the ULA on which the Sinclair ZX Spectrum design was initially based. The RB and RC types by comparison have resistors which are orders of magnitude larger. Migrating to a ULA type that had very different component values would have required a significant redesign and testing effort.

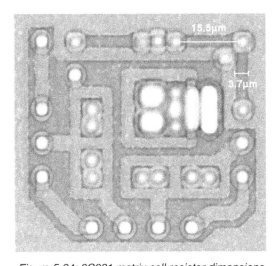

*Figure 5-34: 6C001 matrix cell resistor dimensions*

Figure 5-34 shows an optical image of the 6C001 matrix cell after the metallisation layer has been removed. As the dimensions of all three resistors are identical, their resistance will also be identical. For clarity the length of Rcs and the width of RL2 are shown. Given these dimensions, the resistance of the matrix cell resistors may be calculated as:

$$665 \times (15.5\mu m \div 3.7\mu m) = 2786 ohms$$

This value is close to that of RA matrix cell resistors RL and Rcs, at 2600 and 3000 ohms respectively, again demonstrating its relationship to the RA ULA.

The Sinclair 6C001 ULA makes use of the two independent power rails, one to supply the peripheral cells and the other to supply the matrix cells via the series voltage regulators, Tvs. This is illustrated by Figure 5-17. By being driven from two isolated supplies, the ULA allows current to the matrix cells to be controlled independently of the peripheral cells, reducing power consumption and heat dissipation.

Last we present two scanning electron microscope images of the 6C001 ULA showing how advanced and precise Ferranti's process was, being able to produce metallisation track widths of just 3.2 microns across, plus the depth of the silicon wafer compared to the surface detail.

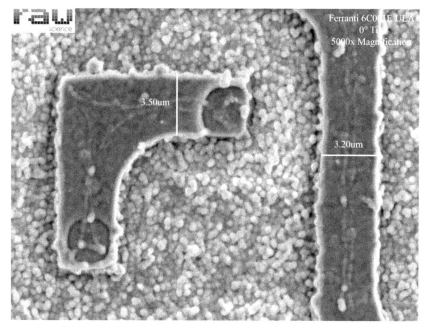

*Figure 5-35: Scanning Electron Microscope image of ULA 6C001 track widths*

Ferranti 6C001E ULA
20° Tilt
250x Magnification

*Figure 5-36: SEM image of ULA 6C001 showing chip and layer depths*

## Package Types

ULAs were available in several packages, as shown in Table 2.4.

| Package Code | Description |
|---|---|
| E | Plastic DIL |
| J | Ceramic DIL |
| G | Plastic FLATPACK |
| F | Ceramic FLATPACK |

*Table 5-5: ULA package types*

---

1. The ULA Technical Manual, 1980 [ULAHAND] defines the 5C000 ULA as containing 440 matrix cells, and the accompanying photo confirms this. A Ferranti marketing publication from 1979 [FERRANTILSI] shows a

large photo of a ZNA5002 ULA whose structure is almost identical to the 5N000 and 5C000 ULA, titled "ULA 5000 committed LSI circuit", but which has 484 matrix cells.

It is interesting that there is no performance identification in the ZNA5002 string (i.e. no C, H, L, N or U), implying that the ULA is a normal, 3MHz maximum device.

The ULA Technical Manual does not describe such a device in either the Product Data Summary, section 5.3 (ULA1000 and ULA2000 being the only standard devices listed), or in the technical product data sheets for each device.

However, further investigation of the ULA Series Numbering System reveals that ULA 50000 devices could have 484 matrix cells: "5 - means 5000 Series - 440 to 484 cells for system complexities up to 1000 gates", indeed, the 5L000 device has a matrix size of $22 \times 22 \times 1$ cells, but this RTL device has a structure entirely different to the ZNA5000 device. Examining a closeup of the ZNA5000 photo shows the matrix cell to contain three resistors, a dual transistor, a single transistor and a current source. The expectation was that this would be an RTL device, and would not contain a current source. The peripheral cells too were unusual in that they contain three or four resistors, and three transistors.

My suspicion is that the 5000 device is in fact an early 5000 series CML ULA, as the 5N000 and 5C000 devices contain four matrix cell transistors and many more components in their peripheral cells. Therefore the 'normal' device type in Table 5-1 may be CML for the 5000 device. See Table 5-2.

# Chapter 6
# Sinclair and the ULA

Sinclair's first home computer, the ZX80, was launched as a kit in February 1980 at a price of just £79.99, equivalent to £227.97 in 2010. To meet this price target its designer Jim Westwood used just 17 off-the-shelf logic chips, 1K of RAM and 4K of ROM, which contained the BASIC interpreter written by Nine Tiles Information Handling. It was simple in design and made clever use of specific Z80 instruction fetch cycles to maintain the output of the television picture.

There was little improvement that could be made to the ZX80 hardware within the £79.99 price tag, so with Sinclair's decision to produce a more advanced follow-up machine, came the need to reduce production costs.

## The First Sinclair ULA

The design of the ZX80's successor, the ZX81, started in the September before the ZX80 was released. The cost associated with the proposed ZX81's larger and more functional ROM, again from Nine Tiles, and its increased hardware complexity meant it would not be commercially viable and meet Clive Sinclair's strict price point unless the chip count could be minimised to reduce costs. By this time Ferranti and their CML ULA were well established in the semi-custom device market, and being UK based were ideally placed to provided Sinclair the means with which to achieve its goal.

Westwood took what was basically a modification of the ZX80 and turned this into a Ferranti ULA 2000 series logic design, which he wire-wrapped into a prototype but had little success in getting to work. In frustration, Westwood was forced to leave it in the hands of the new recruit, Richard Altwasser, while he took a week away from the office on business. Altwasser says he doesn't know who was more surprised when Westwood returned and unexpectedly found him in possession of a working prototype.

From this design Westwood created an interconnect layer for the Ferranti 2C000 ULA, producing the 2C184E and later the 2C210E. Around this he added the same Z80A as the ZX80, the larger 8K of ROM and again 1K RAM, creating a machine with just four ICs.

This drastic reduction in chip count significantly reduced the power consumption that was exhibited in comparison with the ZX80, which reduced the heat output and made the computer much more stable in operation. The lower chip count and manufacturing costs ultimately meant that in 1981, Sinclair could sell the ZX81 in kit form for just £49.99, equivalent to £134.47 in 2010.

With the Ferranti ULA at hand through which to realise their future technical designs, Sinclair had set the stage for the grand entrance of the ZX Spectrum.

*Figure 6-1: Sinclair ZX81 2C210E ULA*

## The ZX Spectrum ULA

Having completed the ZX81, Westwood's experience with television saw him move on to television research and development, leaving the clearly capable Altwasser to head development of the ZX81's successor, the ZX82, as the ZX Spectrum was originally called. Work on the specification for the ZX Spectrum began in September 1981, and was mostly compiled from internal discussions between Altwasser, Westwood and Nine Tiles. There was little question that it should inherit much from the ZX81, notably the TV UHF output and reuse of code from the ZX81 ROM. This meant employing the same Z80 processor, which was also a good cost choice. The design of the ZX Spectrum was further constrained by Clive Sinclair's desire to launch the machine at the IPC Computer Fair at Earls Court on 23 April 1982, as well as his customarily low price target.

Altwasser wrote and agreed the technical specification quite quickly, and very little changed during subsequent development. One of the requirements of the design was for it to feature sound and colour, and it was television text standards that influenced the decision to use a single colour attribute per character which, as a design benefit, reduced the amount of memory required. Clive Sinclair took the decision to use Ferranti CML, having been convinced that this was a superior technical decision for the ZX81, and strongly guided Rick Dickinson's case designs. Overall, cost drove the choice of CML ULA and the memory chips used, and the need for minimal silicon real estate.

Having visited Ferranti to understand the CML ULA technology and its constraints, Altwasser produced the logic design entirely on paper and prototyped it using wire wrapped SSI 74S and 74LS TTL logic chips, all in just a few weeks. The logic design required a good understanding of the capacity and analogue capabilities of the chosen 5000 series ULA, since very few gates would remain unused, and the analogue video output made use of peripheral cells intended by Ferranti for digital interfacing. Compromise had to be made throughout as the ULA did not have the gate capacity to realise a fully synchronous design, and the interface signals had to be kept to a minimum by reducing functionality or through multiplexing to keep the total pin count from exceeding 40. The design made extensive use of the Ferranti component library, providing many of the necessary building blocks such as flip-flops and TTL outputs. Critical paths were identified by design analysis and considering gate loading, and mainly focused on memory access timing and the video output signals. Altwasser defends the lack of computer simulation, arguing that the simplicity of the design meant that the analysis and understanding of all the critical paths was within the grasp of one engineer.

When the logic design and prototype were complete, the placement of functional units within the ULA and layout of the interconnection layer on mylar film was carried out by Ferranti engineers at their offices in Oldham, Manchester. This was done jointly with Altwasser, who occasionally returned to Oldham to make layout decisions, and considered pin-out requirements that simplified PCB layout and minimised high speed signal tracks. Once the interconnection layout was complete, critical path analysis was repeated. Altwasser notes that although all done manually, the steps were entirely in keeping with modern day automated post-routing simulation. He and the Ferranti engineer would study the mylar film and measure track lengths for the critical signals, calculate parasitic capacitances and the resulting signal delays and slew rates, and manually plot timing curves to ensure they were within limits. Where necessary, Altwasser would request track routing changes or even change the logic, perhaps adding buffers to meet the timing constraints.

Altwasser used both the clock and varying amounts of propagation delay to establish the desired signal timings. Because the propagation delay of a matrix cell logic gate is predictable and tunable within a range, he was able to delay and stretch signal pulses with far less complexity than by using synchronous flip-flops or counters.

Having finalised the interconnect layout, it was optically reduced to a $10\times$ magnification light-field mask and used to create a complete multi-chip wafer mask using an optical step-and-repeat procedure, as discussed in the section called *Generate Interconnect Pattern* in Chapter 5, *The Ferranti ULA*. From this Ferranti produced an initial batch of prototype wafers and invited Altwasser to functionally test and verify the chips in-situ, before they were diced and packaged.

Altwasser was provided with a wafer probe that allowed connection to individual bond pads and made it possible to attach external circuitry with which to test the devices. Altwasser and Ferranti were under extreme time pressure, and carrying out tests during this stage saved a few precious days. However, while performing these tests and visually inspecting the interconnection tracks under a microscope, it was discovered that the Ferranti layout engineer had made an error in the interconnect layer whereby the clock output from the early counter stages was not connected to the later stages. Fortunately, and against all odds, a tiny fleck of dust had fallen onto one of the devices of the multi-chip wafer mask at exactly the point at which the missing interconnect should have been, despite the usual clean room conditions of a semiconductor plant. This artificial bridge prevented etching of the aluminium at that point, connecting the clock to the later counter stages and allowing Altwasser to complete his full test suite on this one die, successfully proving the entire chip design of the 5C102E ZX Spectrum ULA.

The first two issues of the ULA were based on the largest version of the array used for the ZX81, which provided twice the number of gates. The later issues of the ULA used a new, and larger still, 6000 series array, which in the main provided a reduced power consumption.

# Chapter 7
# The ZX Spectrum Overview

Before looking at the ZX Spectrum ULA in any detail it is useful to consider the design of the ZX Spectrum at a high level, introducing the main functional units and the relationship between them. These may be summarized as follows:

1. Z80A CPU
2. 16K ROM
3. 16K RAM
4. 32K RAM (Optional)
5. Clock Generator
6. Colour Video Generator at a resolution of 256 × 192
7. Colour Encoder
8. Keyboard Input Port
9. Cassette I/O Ports
10. Internal Speaker

The complexity of these units varies, some containing just simple interface logic, whilst others are made up of complicated state machines and a large number of precisely timed signals.

## The Z80A CPU

Even though the Zilog Z80A CPU is in itself the most complicated component in the ZX Spectrum, its interface requirements are simple as it requires nothing more than a clock signal of up to 4MHz and some memory, for which it provides an address bus, a data bus and read and write control signals. In the ZX Spectrum, the Z80 is driven by a 3.5MHz clock, generated by the ULA.

## 16K ROM

The program stored in the ROM is the work of John Grant and Steven Vickers, contracted from Nine Tiles Information Handling Ltd. The program contained within the ROM cannot be changed, and is the first thing that the Z80 executes when it is switched on. The familiar "© 1982 Sinclair Research Ltd" message is generated by the ROM.

Most importantly, the ROM contains the necessary code to get the ZX Spectrum's hardware to do something useful. It monitors the cassette interface with such precision that it is able to differentiate between binary 1s and 0s, essential if programs are to be loaded into memory. It scans the keyboard 50 times a second and works out what key combinations have been pressed, and it contains code that will write the correct sequence of bytes to the video display memory to have the TV display strings of text.

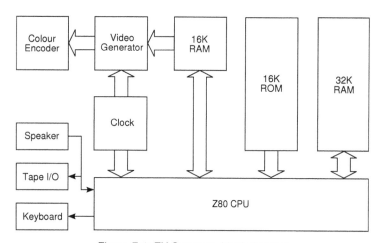

*Figure 7-1: ZX Spectrum block diagram*

## 16K RAM

The 16K of RAM present in both 16K and 48K models of the ZX Spectrum is used by the ROM to store important system variables and user programs. A section of this memory is reserved as the video display memory, which is read by the Video Generator 50 times a second in Europe (60 times a second in the U.S. and Canada.

The 16K RAM is of the dynamic variety and consists of eight 1-bit 4116 DRAM chips [DS4116]. Internally, the 4116 16K RAM devices contain 16384 locations, organized as a matrix of 128 rows by 128 columns and fed by a 7-bit multiplexed address bus. The Z80 CPU does not have a multiplexed address bus itself, so the ZX Spectrum splits the fourteen CPU address lines required for 16K memory access into two 7-bit addresses with a pair of 4-bit multiplexers, and then routed to the dynamic RAM. Which of the 7-bit addresses is presented to the DRAM at any one time is determined by one of the many dynamic RAM control signals generated by the ULA. See Chapter 13, *Video Memory Access*, and Chapter 17, *CPU Memory Access*, for further details.

## 32K RAM

The upper 32K of RAM is present in 48K models of the ZX Spectrum, and was available as an upgrade to 16K models. As with the 16K RAM, this memory is also of the dynamic variety and requires a multiplexed address bus, provided by two additional 4-bit multiplexers and some control logic not present in 16K models. The RAM itself consists of eight 1-bit 4532 DRAM chips.

## Clock Generator

The clock generator is a sub-component of the ZX Spectrum. It is the 14MHz master clock signal from which all the other timing signals used within the ZX Spectrum are derived. The master clock is divided by two to provide the 7MHz pixel clock, and divided by four to provide the 3.5MHz CPU clock.

## Video Generator

The video generator reads the first 7K or so of the 16K RAM, and generates a video signal at a resolution of 256 pixels wide and 192 pixels high containing the display information it found there. The video output consists of a luminance and synchronisation signal, Y, and two colour difference signals, U and V.

## Colour Encoder

The colour encoder takes the Y, U and V video signals from the video generator and combines them into a single composite video signal. The colour encoding may be PAL or NTSC, depending on the local video standard in use, and is set at the time of manufacture.

## Keyboard Input Port

The keyboard input port samples the keyboard matrix and calculates what combination of keys are been pressed.

## Cassette I/O Ports

The cassette I/O ports are analogue circuits which associate voltages with the binary states 0 and 1. A very low, or zero, voltage is associated with state 0, and a higher positive voltage with state 1. These voltages are sent to and received from an external cassette recorder, and forms the basis of program and data storage. The ROM program processes the data it is writing to cassette with a clever algorithm to determine the exact sequence of high and low voltages needed to unambiguously represent each data byte. It does this to ensure that the data may be accurately read back from cassette at a later date.

## Internal Speaker

The speaker is a small, high impedance speaker whose coil may be energized by sending a binary 1 to the output port it is connected to, and deactivated by sending a binary 0. Alternating between binary 1 and 0 at a particular frequency causes the speaker to oscillate at that frequency, and produce the related tone.

## The ULA Chip

The ULA is the core of the ZX Spectrum. It performs the role of video generator, CPU clock generator, memory access governor, keyboard controller, cassette I/O and speaker controller. It coordinates CPU access to these resources so that the television display is never interrupted, and performs the necessary conversion between the analogue television and interface signals and the digital signals used by the processor.

How the ULA achieves all this is explained in the following chapters.

# Chapter 8
# The Memory Map

The ZX Spectrum employs the Zilog Z80 CPU, an 8-bit microprocessor having an 8-bit wide data bus and a 16-bit wide address bus. It therefore transfers eight bits at a time from any one of 65536 memory locations ($2^{16}$ = 65536).

The Z80 reads from memory the instructions that it is to execute, one byte at a time, starting at memory address zero. After each byte fetch the Z80 increments its 'current memory location' address by one, and reads the next byte. Without a program to execute, the Z80 does not know what to do and becomes nothing more than a useless collection of transistors.

To avoid this, computers provide a built in program that immediately gives the processor something constructive to do. In this respect, the ZX Spectrum contains a program written by Steven Vickers and John Grant that initialises its hardware, sets all of the available RAM locations to contain the byte zero, clears the screen and displays the familiar copyright message before going on to perform other tasks. For such an initialisation process to be reliable, the first instruction of this program must be located at memory address zero, so that it is the first thing that the Z80 executes when its power is applied.

Normally memory forgets whatever it was storing when power is removed, so to ensure that the initialisation program is always available at power on, it is stored in non volatile ROM. This has led to the program being affectionately called "The ROM", as featured in the title of the Dr Ian Logan and Dr. Frank O'Hara book "The Complete Spectrum ROM Disassembly", although they refer to the program within the book's opening pages as "the monitor". The ROM program is 16 kilobytes in length, and extends from address location 0 through to 16383.

The remainder of the ZX Spectrum's memory map contains RAM. This starts at address 16384 and continues to 32767 in the 16K model, 65535 in the 48K model. This memory is used to store the video display information, temporary variable storage for the ROM program, user programs and data.

The complete memory map of a ZX Spectrum is shown in Figure 8-1.

The lower 16K of RAM is provided by a dedicated bank of RAM chips and is present in all models of the ZX Spectrum, and the first 6912 bytes of this are reserved for and used by the ZX Spectrum's video display circuitry.

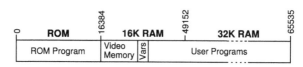

*Figure 8-1: ZX Spectrum 48K memory map*

The upper 32K of RAM is provided by its own dedicated set of RAM chips, which are only present in 48K models.

As there are three distinct memory devices connected to the Z80 (ROM, 16K RAM and 32K RAM), the ZX Spectrum must ensure that the appropriate one is selected for the memory address required. Memory devices provide an enable signal for this purpose, which are activated by the ULA, depending on what location the CPU is addressing. If the upper 32K is being accessed the ULA does nothing, as this memory is enabled by on-board logic associated with the 32K RAM.

No other memory mapping is carried out by the 48K ZX Spectrum[1] as the ROM, 16K and 32K RAM fill the entire 64K memory space accessible by the Z80 CPU.

---

1.  The ZX Spectrum 128, +2 and so forth divide 128K into 16K pages, and with additional logic allows up to four pages to be mapped into the 64K address space.

# Chapter 9
# The Video Display

## Television and Display Basics

A television works by scanning an electron beam horizontally across the screen. When it reaches the right hand side, it is blanked and flies back to the left and descends by one line, where the process repeats. When it reaches the bottom of the screen, it is blanked and sent back to the top. The picture is created by varying the intensity of the electron beam as it scans the screen.

In a UK PAL television receiver the electron beam takes a fixed time of $64\mu s$ to travel from the left of the screen to the right and back to the start of the next line, and a complete PAL picture frame contains approximately 312 such scan-lines. At a horizontal scan period of $64\mu s$, 15625 scan-lines will be drawn every second, and at 312 lines per frame, the screen will be updated 50 times in that second.

To be used as a computer display the television screen must be divided into a grid of pixels. The number of vertical pixels may not exceed the number of scan-lines in the display, and the number of horizontal pixels that may be achieved is determined by the rate at which the electron beam intensity is adjusted for each pixel; that is, the frequency of the pixel clock. In order for pixels to appear square, the horizontal to vertical ratio should be as close to 4:3 as possible, the ratio of the dimensions of a television tube.

As the television has no memory, it requires a continuous stream of pixel information to keep the display stable, so either the CPU must divert a great deal of its time to this task, or dedicated electronics provided to keep the supply of display information flowing to the television. In either case the display pixel patterns must be stored in memory, from where they are fetched by the CPU or video electronics when required. Using dedicated electronics allows the CPU to get on with other tasks and increases the resolution that is possible with the CPU alone, as higher screen resolutions require faster CPUs.

## Choosing a Display Resolution

If a scan-line were divided into 1000 pixels, they would have to be sent to the television at a frequency of 15.625MHz. If the CPU were used to update the video directly, it would have to operate in excess of this frequency and have little time to do anything else. Such a method of television update would therefore be inappropriate for a high resolution display. Even with dedicated electronics, where a pixel clock of this frequency is quite achievable, the amount of video memory required to hold a display of this resolution would be excessive. For instance, assuming one bit of memory per pixel, then at 1000 pixels a single scan-line would require 125 bytes of memory to be stored, which at a full vertical television resolution of approximately 312 lines, the whole display would occupy more than 38K of RAM. This is more than half the memory available to an 8-bit microprocessor, and the high pixel clock would demand expensive fast memory.

Clearly a screen resolution consuming nearly all of the ZX Spectrum's RAM would not have been an option, and Altwasser's solution is two-fold in that they use a lower frequency for the pixel clock and reduce the display area size by introducing a non-display border around a central pixel display rectangle, concepts inherited from the earlier ZX80 and ZX81.

The ZX81 had an effective display resolution of $256 \times 192$ pixels, but because of its limited 1K of RAM, was essentially character ROM driven and had no video memory at this granularity. Instead, the display was divided into a grid of $32 \times 24$ character cells, with 768 bytes of memory reserved to store the ASCII code of the character at each cell position. A simple and cheap hybrid video system was used where the Z80 CPU played a part in the display update by reading the pixel pattern for each cell from a character ROM, and sent them to the video output circuitry at the appropriate time.

The ZX Spectrum adopts the same resolution as the ZX81, but assigns one bit of display memory to each pixel, so that $256 \div 8 = 32$ bytes are required per scan line, and a total of $32 \times 192 = 6144$ bytes for the entire screen. It has dedicated electronics to maintain the video display, relieving the CPU from this responsibility.

## Vertical Interlace

A broadcast UK PAL television picture actually consists of 625 interlaced scan lines split between two frames of 312.5 lines each. The frames are alternately displayed 50 times a second, with the second vertically offset by half a scan line so that it fills the gaps between the lines of the first frame. These frames are referred to as frame 1 and frame 2. Due to persistence of vision, and the relatively slow emission decay of phosphorus, interlacing has the effect of doubling the vertical resolution without increasing the bandwidth per frame.

Producing a stable interlaced computer display is difficult as interlacing tends to produce a visible vertical jitter, so most machines, including the ZX Spectrum, only generate a single 312 line frame 50 times a second. A single frame should contain 312.5 lines, however as half lines are of little use, it is usual for just the whole lines to be used.

## Positioning the Display

The 192 lines of the ZX Spectrum's pixel display rectangle are vertically aligned centrally on the screen creating a visible top border of 56 lines. This is preceded by eight invisible lines during which the electron beam is in its vertical fly-back. The remaining 56 lines provide the bottom border and an invisible off-screen region.

Once an appropriate memory-conservative screen resolution had been agreed, a suitable pixel clock frequency needed to be established to maintain a 4:3 display ratio. If it were too high, the pixels would be drawn on the screen before the electron beam had moved very far, creating a display that would be squashed over to the left hand side. If the clock were too low, the electron beam would reach the right hand side of the screen before all the pixels had been displayed, creating a display that was wider than the width of the television.

Some of the scan-line lies outside the visible portion of the screen, and approximately $6.9\mu s$ is taken by the electron beam returning to the left hand side of the screen. Similarly not all scan-lines are visible, as some exist above and below the visible screen area including the vertical fly-back.

To determine the pixel clock that would create a square pixel, some calculations are required. As the horizontal to vertical ratio of 256:192 matches the screen aspect ratio of 4:3 the scaling factor is 1:1. Therefore the percentage of a scan-line used horizontally for pixel display will equal the percentage of scan-lines used vertically.

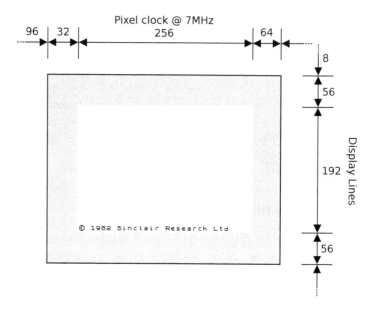

*Figure 9-1: PAL horizontal and vertical screen dimensions*

First the percentage of non vertical retrace scan-lines that are used for pixel display is calculated:

$$\frac{192}{(312-8)} = 63.158\%$$

Second the same percentage of the 64$\mu$s scan period, minus the horizontal flyback time (figures from Chapter 11, *Video Synchronisation*), is calculated

$$\frac{(64-(13.7-(2.2+4.6)))\mu s}{100} \times 63.158 = 36.0632\mu s$$

And finally the frequency at which 256 cycles take 36.0632$\mu$s:

$$\frac{256}{36.0632\mu s} = 7.09 MHz$$

At a pixel clock of 7MHz, the time to display a row of pixels is 36.56$\mu$s or 57.14% of the total scan-line period. On an average television this pixel area takes up approximately 74% of the visible screen width, allowing for left and right borders of 22% including the non visible area. As expected, these proportions give a display dimension that produces square pixels, is aesthetic and fits comfortably within the screen.

At this pixel clock there are 448 cycles in a 64$\mu s$ scan line period, and a pixel display row consumes 256 of these, one for each pixel. Due to design simplifications in the ZX Spectrum the pixel display is offset towards the left hand side of the screen, as the left and right borders are slightly different widths of 32 and 64 cycles respectively. The remaining 96 cycles are used off-screen and during the horizontal fly-back. See Figure 9-1.

## The NTSC Display

The NTSC specification defines a television frame as containing 525 scan lines, split across two interlaced fields; therefore a single field contains 262.5 scan lines. As an NTSC scan line takes 63.55$\mu s$ to cross the screen, the frame rate is approximately 60Hz.

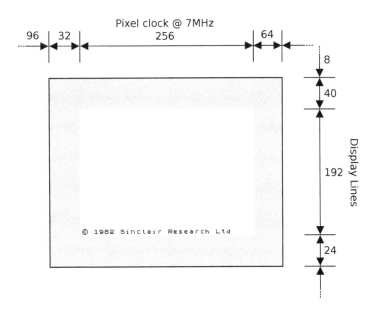

*Figure 9-2: NTSC horizontal and vertical screen dimensions*

85

By using 264 scan lines of 63.5$\mu$s each, a frame rate of 59.65Hz is produced. This is acceptably close to the NTSC specification, and allowed the Sinclair engineers to use same horizontal control circuits in both PAL and NTSC versions. See Figure 9-2 for the ZX Spectrum NTSC screen dimensions, and the section called *The NTSC Line Counter* in Chapter 10, *The Internal Clocks* for details of the NTSC scan line period.

# Chapter 10
# The Internal Clocks

The primary task of the ULA is to keep the video display updated and as this is such as precise and exacting requirement as far as timing is concerned, the internal state machine provides video orientated time events, from which all other system states and time points are derived.

## The Oscillator

At the heart of the ULA is its crystal controlled oscillator. Because the ULA does not contain any capacitors, the oscillator design departs from convention and uses transistors in an inherently unstable configuration to achieve oscillation, the resonant frequency of which is controlled by an external crystal.

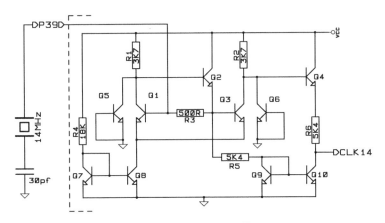

*Figure 10-1: 14 MHz oscillator*

The basic operation of Figure 10-1 is as follows. Two transistors, Q1 and Q2, are configured to oscillate, the frequency of which is moderated by the 14

MHz crystal. The sine wave produced is then amplified and processed by Q3 and Q4 into a clean square wave.

With Q1 initially off, its collector and the base of Q2 will be at Vcc. Current will flow in the base of Q2 switching it on, raising the voltage at its emitter to approximately Vcc. The current now flowing through Q2 and R3 causes Q1 to turn on, lowering the potential at Q1's collector which reduces the current flowing through Q2's base, causing it to switch off. This reduces the voltage at Q2's emitter and therefore the current in Q1's base, switching off Q1. The cycle then repeats.

The crystal placed between the base of Q1 and ground tunes the oscillation to its 14MHz resonance frequency, R3 buffering the crystal from the signal output at the emitter of Q2.

The collector-follower Q3 provides some gain and inverts the oscillating signal, feeding the output emitter-follower Q4. One would expect the final signal output to be taken from the emitter of Q4, its point of highest impedance, but instead it is taken after the emitter resistor R6. This is to allow some signal shaping to be performed, as a square wave and not a sine wave is desired. This action will become apparent after we have discussed the current mirrors Q7–10.

Q7 and Q8 are configured as a classic current mirror, in that the current flowing in the emitter of Q8 will mirror the current flowing in the emitter of Q7, which is set by R4. Thus R4 indirectly controls the maximum current flowing in the emitters of Q1 and Q3.

Q9 and Q10 are similarly configured as a current mirror, however the current being mirrored is not constant as it is modified by the current flowing in the emitter of Q2. This is how the signal shaping into a square wave is applied.

When Q2 is on, the maximum current allowed by R5 flows in the emitter of Q9. Q3 will be on, holding Q4 off. The output CLK14 therefore goes low and sinks up to the mirrored current through Q10. When Q2 is off, no current flows in the current mirror. Q3 is also off, leaving R2 to pull Q4 on. As no current can now flow through Q10 of the current mirror, CLK14 goes high by sourcing current through R6, which is why the output is taken after emitter resistor R6 and not before. As R5 and R6 have the same value the current sourced or sunk by CLK14 will be the same. This ensures that the positive and negative edges of the clock pulse are symmetrical, as a transistor's switching speed is related to the current flowing through it.

The remaining two transistors Q5 and Q6 are configured as reverse biased diodes connected between the collectors of Q1, Q3 and ground to introduce some temperature stability.

This circuit is implemented by peripheral cells 21 and 22, the former being connected to the crystal via pin 39.

## The 7MHz Clock

By driving the ULA with a master clock at twice the required frequency and dividing internally to 7MHz, Altwasser achieved an accurate and stable clock at the required matrix cell voltage of Vs, and any frequency drift or duty cycle irregularities in the master clock are halved. The inverter on the output of the 7MHz clock divider in Figure 10-2 is configured to have a large fan-out and fast switching speed in order to drive the many circuits that require the clock, without any deterioration in the shape of the signal.

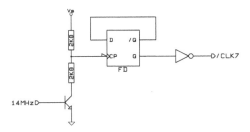

*Figure 10-2: 7MHz clock*

This high-drive /CLK7 signal is routed to the multitude of circuits that require it. Routing the inverted clock allows circuits to either invert it back to CLK7 or NOR gate it with other signals and still remain synchronous with each other. /CLK7 is never used directly. Figure 10-3 illustrates this, showing a gate in line with /CLK7 before each flip-flop.

All ULA timing states are synchronized to this internal 7MHz clock, and through its division all other clocks are generated.

## The Master Counter

As discussed previously, Altwasser's choice of 7MHz pixel clock produces 448 cycles in a 64$\mu$s display scan line period. The master horizontal counter is configured to count through these 448 distinct states, providing the timing reference for display control signals such as the horizontal sync and all other internal system signals.

The counter is split into two stages, the first as a free running 6-bit counter consisting of negative edge triggered D-type flip-flops (see Appendix B, *Component Library*), the most significant bit of which, /C5, clocks the second stage, a 3-bit synchronous counter of negative edge triggered T-type flip-flops with reset, carry and enable.

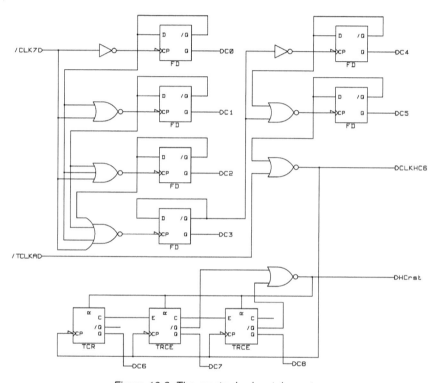

*Figure 10-3: The master horizontal counter*

The two stage counter is a compromise between the ideal design and conservation of space. The simplest text-book counter consists of D-type flip-flops configured to toggle between 0 and 1 at each clock pulse. The output of each such flip-flop feeds the clock input of the next so that when it switches from 0 to 1 and back to 0, it causes the following flip-flop to toggle. These are called ripple-counters, referring to the propagation of the clock along the chain of flip-flops. The disadvantage with these counters is the delay incurred between the source clock and each flip-flop switching, which doubles for each successive bit.

Such counters are undesirable for state machines, as the propagation delay ripple across the counter bits can cause glitches in the subsequently driven logic. What is sought is a synchronous counter where all flip-flops switch simultaneously on a common clock.

Synchronous counters are more complicated than ripple-counters and consume more components, so the limited space within the ULA forced Altwasser to strike a balance between complexity and gate count by allowing the clock to ripple between just two of the eighteen counter flip-flops.

The first stage is a 6-bit counter consisting of D-type flip-flops that are, all but one, connected in a synchronous configuration, such that each bit clocks only when all the less significant bits are high. Each flip-flop is therefore clocked by the common clock, in this case the 7MHz CLK7 signal, which is NOR gated with the invert of the preceding bits. This makes conservative use of matrix cells, requiring only three for a D-type flip-flop and between half and one cell for the clock gate[1].

However, as each successive bit of the counter uses an additional clock gate input, the number of matrix cells required increases every four bits. To avoid using an increasing number of matrix cells, Altwasser breaks the synchronism between C3 and C4 and allow C3 to ripple forward as the clock of bits C4 and C5, which are configured to be synchronous with this new clock; thus bits C4 and C5 are synchronous with respect to each other, but asynchronous with respect to C3–0, and clock around 24ns later. The final bit C5 is gated by /TCLKA, discussed in Chapter 23, *Hidden Features and Errors*, which may always be assumed to be low, producing clock CLKHC6 which drives the second counter stage. This clock delayed with respect to CLK7 as it transitions around 48ns later.

| State | C8 – 0 |
|---|---|
| 0 | 000 000000 |
| 31 | 000 111111 |
| 32 | 001 000000 |
| 447 | 110 111111 |
| 448 | 000 000000 |

*Table 10-1: Horizontal clock states*

91

The second counter stage is a true 3-bit synchronous counter which is reset immediately on reaching binary 111. As the first counter stage counts from 0 through to 31, the second stage counts from 0 through to 6 to give 32 × 7 = 448 states in total; 0 to 447. Table 10-1 shows these states in terms of their binary counter bits C8 to C0. As the T-type flip-flops have synchronous reset, the reset is generated and held while C7 and C8 are high, forcing a reset of this stage at the next clock transition which, had it not been for the reset, would have advanced the counter to 7 (111 binary).

There is no reset required for bits C5–0, as they will already be at zero when the reset of C8–6 occurs.

## The Vertical Line Counter

### The PAL Line Counter

In addition to keeping track of how far through a scan-line it is, the ZX Spectrum ULA also keeps track of how many scan-line periods have occurred and therefore which line it is currently displaying.

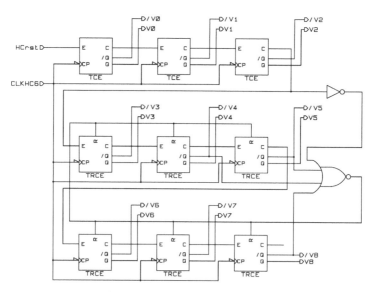

*Figure 10-4: PAL vertical line counter*

This vertical line counter is a 9-bit synchronous counter utilising the carry and enable inputs of the T-type flip-flops from which it is constructed. The flip-flops share a common clock, which is the same as that driving the second stage of the master horizontal counter, making the line counter synchronous with the horizontal counter second stage.

The enable input of the least significant counter bit acts as the overall count enable for the counter, and is held disabled until the horizontal counter reaches the end of its sequence. Recall that the horizontal counter's second stage synchronous reset, HCrst, is activated one CLKHC6 period prior to the desired reset point, so that the second stage is immediately reset at the next CLKHC6 transition, as the counter advances to 448. Since the vertical counter is also clocked by CLKHC6, controlling the counter enable with HCrst ensures that the vertical counter increments at the exact moment that the horizontal counter resets to zero.

The PAL variant of the ZX Spectrum ULA develops 312 scan lines per television frame, and so the vertical line counter is reset as it advances from 311 (100110111 binary). The upper six flip-flops of the counter therefore have an additional synchronous reset for this purpose. As the reset is synchronous with the clock, it is applied while the counter reads 311 so that the reset occurs at the next clock transition when the counter steps to 312; thus the reset is generated from the NOR of /V8, /V5, /V4, /V2–0:

$$VRst = \overline{\overline{V8} + \overline{V5} + \overline{V4} + \overline{V2} + \overline{V1} + \overline{V0}}$$

Interestingly, instead of using the three least significant counter lines V2–0, Altwasser uses the invert of the carry from V2, which is high whenever V2–0 are high, as is the case when the counter prepares to advance from 311.

This greatly simplifies the reset logic, resulting in the following equations, as implemented in Figure 10-4:

$$VRst = \overline{\overline{VCA_2} + \overline{V4} + \overline{V5} + \overline{V8}}$$

### The NTSC Line Counter

The NTSC version of the ZX Spectrum ULA is clocked by a slightly faster 14.11 MHz crystal. This shortens the time taken for a complete cycle of the master counter, and thus scan line, to $63.5\mu s$, complying with the NTSC specification.

The NTSC ULA generates 264 scan lines per television frame, and resets the vertical line counter as it advances from 263 (100000111 binary). This produces the NTSC frame rate of 59.65 frames per second.

As described by the section called *The PAL Line Counter*, the vertical counter is synchronously reset. Therefore, by applying the reset signal while the counter reads 263, the counter is reset as it advances to 264.

The differences between the PAL vertical counter reset of 100110111 and the NTSC counter reset of 100000111 are at bits V4 and V5, leaving V8 and V2-0; therefore the reset signal of the NTSC ULA contains only $/VCA_2$ and $/V8$:

$$VRst = \overline{VCA_2 + V8}$$

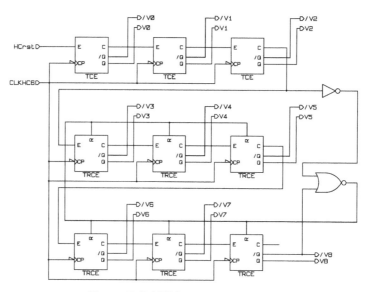

*Figure 10-5: NTSC vertical line counter*

---

1. For comparison, the alternative synchronous counter consisting of T-type flip-flops would require five matrix cells per bit.

# Chapter 11
# Video Synchronisation

To produce a display on a television, the motion of the electron beam must be synchronized to the ULA state machine. To achieve this the video feed must contain two synchronization signals: the horizontal sync that sends the electron beam back to the left, and the vertical sync that returns the beam to the top.

## Horizontal Timing

The PAL standard specifies that during the horizontal sync and fly-back, the video signal must be blanked for a period of at least $12\mu s$. The horizontal sync pulse that initiates the fly-back is $4.7\mu s$ in duration and occurs within the blank period, after approximately $1.5\mu s$ of front-porch delay.

Figure 11-1 shows the ZX Spectrum video components, with timing given in microseconds and the associated pixel clock states. Each period is derived from multiples of 16 clock states, or $2.29\mu s$. The blank and front-porch periods are therefore slightly longer than standard, at $13.7\mu s$ and $2.29\mu s$ respectively.

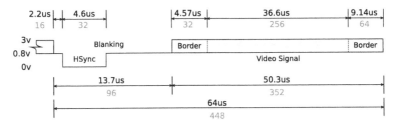

*Figure 11-1: ZX Spectrum horizontal timing and clock states*

The front-porch timing differs between revisions of the ZX Spectrum, the 5C ULA found in issue 1 and 2 machines has a front-porch of $2.2\mu s$, the 6C ULA found in later issues has a front-porch of $3.4\mu s$. This longer front-porch and consequently later horizontal sync has the effect of shifting the display slightly further to the left of the screen, as the time between the sync and the start of the video output is thus shorter. Note also, that the left and right hand borders are of different widths, contributing to the display being offset to the left.

| Description | Cycle Start | Cycle End | C8–0 at Start |
|---|---|---|---|
| Pixel Output | 0 | 255 | 000 000 000 |
| Right Border | 256 | 319 | 100 000 000 |
| Blanking Period | 320 | 415 | 101 000 000 |
| Horizontal Sync | 336 (5C) | 367 (5C) | 101 010 000 (5C) |
| | 344 (6C) | 375 (6C) | 101 011 000 (6C) |
| Left Border | 416 | 447 | 110 100 000 |
| Synchronous Counter Reset | 447 | 448 | 110 111 111 |

*Table 11-1: Horizontal time points for the 5C and 6C ULA*

The exact division of the horizontal scan line in terms of master counter states is given by Table 11-1. Counter states 0 to 255 define the region of pixel display, such that C8 may be used to determine whether pixels or right-to-left border colours are to be output. During the horizontal sync, the blanking period prevents any video output.

## Horizontal Blanking

The blanking period of $13.7\mu s$ is enabled when C8–5 equals 1010, and continues until it reaches 1101. C8 is the active-low pixel enable signal and thus will be high during the horizontal blank. The most significant bits C8–6 of the period remain at 101 for $9.1\mu s$, after which they become 110 with C5 low for a further $4.6\mu s$. The combination of the high state of C6 followed by the low state of C5, selected by considering C8 and C7, generates the $13.7\mu s$ blanking period. The two horizontal blank components are shown visually in Figure 11-2, and are defined by:

$$Blank1 = \overline{\overline{C8} + C7 + \overline{C6}}$$
$$Blank2 = \overline{\overline{C8} + \overline{C7} + C5}$$

Giving /HBlank as:

$$\overline{HBlank} = \overline{Blank1 + Blank2}$$

## Horizontal Synchronization

The horizontal sync pulse occurs within the blanking period, after a front-porch delay of $2.29\mu s$ or $3.43\mu s$, depending on ULA version. A number of internal signals are generated, the combination of which produces the $4.6\mu s$ horizontal synchronisation pulse.

First a train of horizontal sync pulses is generated, delayed with respect to the horizontal blank start to give the desired front-porch period.

C5 forms the basis of the horizontal sync pulse as it changes state every $4.6\mu s$. However, without any further treatment it is aligned to the horizontal blank and gives a front-porch of 0 or $4.6\mu s$, depending on which half cycle of C5 is taken.

The 5Cxxx ULA uses half the period of C4, at $2.29\mu s$, to set the length of the front-porch. C5 is modified by C4 to create two signals, the first contains pulses derived from the high state of C4 when C5 is low, the second from the low state of C4 when C5 is high. When combined these two signals produce a train of pulses of $4.6\mu s$, offset by $2.29\mu s$ with respect to the clock.

From Figure 11-2 and Table 11-1 it may be shown that the horizontal sync of the 5Cxxx ULA is given by:

$$\overline{HSyncA} = \overline{C5 + C4}$$
$$\overline{HSyncB} = \overline{\overline{C5} + \overline{C4}}$$

$$\overline{HSyncPulses} = HSyncA + HSyncB$$

The later 6C001 ULA increases the front-porch slightly to $3.43\mu s$ by introducing C3 into its production, adjusting the pulse alignment by $1.14\mu s$.

$$\overline{HSyncA} = \overline{C5 + X}$$
$$\overline{HSyncB} = \overline{\overline{C5} + \overline{X}}$$

Where:

$$X = \overline{\overline{C4} + \overline{C3}}$$

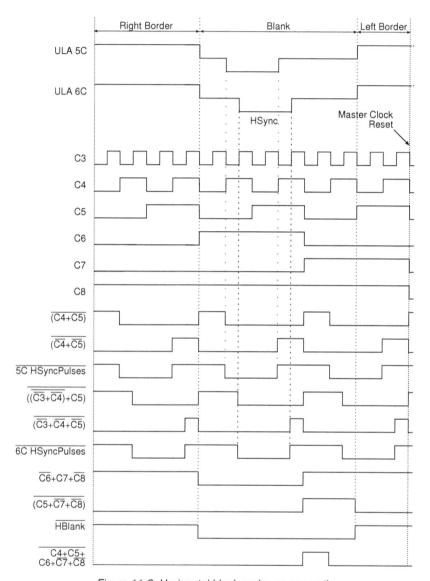

*Figure 11-2: Horizontal blank and sync generation*

Figure 11-3 shows the horizontal blank and sync captured from a 6C001 ULA. Note the glitch in the sync pulse, $4.6\mu s$ from the start of the blank period, which is an artefact of the master counter. As discussed in Chapter 10, *The*

*Internal Clocks*, C3–0 are synchronous with the pixel clock and C3 ripples forward to clock C4 and C5, which are synchronous with each other. C3 therefore goes low approximately 24ns before C4 and C5 go low and high respectively, causing a momentary glitch in HSync, $1.14\mu s$ after its start.

The appropriate horizontal sync pulse is selected from the train of pulses by the first part of the horizontal blank signal, comprised of C6, C7 and C8 such that:

$$\overline{HSyncSelect} = \overline{C8} + C7 + \overline{C6}$$

giving:

$$HSync = \overline{\overline{HSyncSelect} + \overline{HSyncPulses}}$$

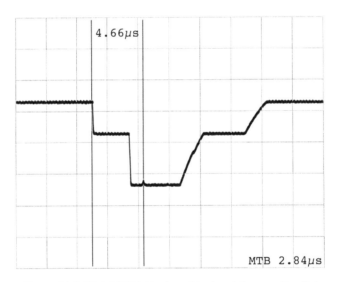

*Figure 11-3: ULA 6C001 blank and horizontal sync with glitch*

## Vertical Timing

Referring to Figure 9-1 we see that the ZX Spectrum's PAL display is split into four vertical sections, comprising 56 lines of top border, 192 lines of pixel display, 56 lines of bottom border and 8 lines of vertical sync period. These periods are broken down into the vertical counter states shown in Table 11-2, where *Lines* represent the internal counter values and not the actual scan lines, such that counter line 0 is the first pixel display row at scan line 56. Scan line 0 begins at the vertical sync pulse.

The main function of the line counter is to generate the vertical sync pulse precisely when required, and for the appropriate duration. In addition it determines whether the ZX Spectrum should be generating pixel display lines, or the top and bottom borders. Referring to Table 11-2 it is clear that whenever (V6 • V7) + V8 is false then pixel display rows are generated, otherwise the ZX Spectrum generates vertical borders or vertical sync.

| Block Description | Lines | Length | V8–0 at Block Start |
|---|---|---|---|
| Display | 000 – 191 | 192 | 0 00 000 000 |
| Bottom Border | 192 – 247 | 56 | 0 11 000 000 |
| Sync Period | 248 – 255 | 8 | 0 11 111 000 |
| Sync Pulse | 248 – 251 | 4 | 0 11 111 000 |
| Top Border | 256 – 311 | 56 | 1 00 000 000 |
| Clock Reset | 312 – 312 | 0 | 1 00 111 000 |

Table 11-2: PAL vertical counter states and associated display regions

The ZX Spectrum ULA logic defines the vertical border as:

$$VBorderLower = \overline{(\overline{V7} + \overline{V6})}$$

$$VBorderUpper = V8$$

Giving:

$$\overline{VBorder} = \overline{VBorderLower + V8}$$

This is combined with the horizontal border signal, C8, into a single border signal that is low whenever the electron beam is within the border or during synchronisation:

$$\overline{Border} = \overline{VBorderLower + V8 + C8}$$

## Vertical Synchronization

A PAL television controls the vertical position of its electron beam with a 50Hz (60Hz for NTSC) saw-tooth oscillator. At the beginning of the oscillator cycle, when its output is maximum, the electron beam is at the top of the screen. As its output drops the beam descends until, at the end of the cycle, the electron beam reaches the bottom. The cycle then repeats. Video signals contain a vertical synchronization component that resets this oscillator at the start of each frame, guaranteeing that the vertical position of the electron beam is synchronized with the incoming video lines.

The PAL and NTSC specifications describe two vertical sync pulse sequences to be used for interlaced video signals. The first sequence is used when preparing an odd numbered video frame, or field 1, the other is for even numbered frames.

Figure 11-4 compares these two pulse sequences. Field 1 synchronization consists of a 5–5–5 sequence of five short pre-equalising pulses, five long vertical sync pulses and five short post-equalising pulses. A short pulse is low for $2\mu s$ followed by a $30\mu s$ delay, and a long pulse is low for $30\mu s$ followed by a $2\mu s$ delay; therefore there are two vertical sync pulses per scan line. See Figure 11-4 sequence A.

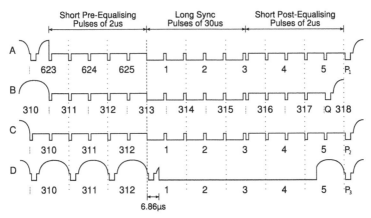

A. Interlaced Frame Field 1 as HSync aligned to VSync (P1).
B. Interlaced Frame Field 2 as HSync 180° out of phase with VSync (Q).
C. Non-Interlaced, single Frame, HSync aligned to VSync (P2).
D. Sinclair Non-Interlaced, single Frame, HSync aligned to VSync (P3).

*Figure 11-4: Comparison of Sinclair and specification vertical synchronisation*

Field 2 synchronization is a 5–5–4 sequence but similar in all other respects to field 1. Being one pulse shorter, the HSync is brought forward by half a line and becomes 180 degrees out of phase with the start of the long VSync pulses. This forces the scan lines to be offset vertically, and thus a field 2 frame. See Figure 11-4 sequence B.

Sequence C shows a general non-interlaced single frame vertical sync. It is a 6–5–5 sequence containing an additional pre-equalising pulse so that a complete number of lines are used for the synchronization, but is otherwise identical to that of field 1. The important similarity of the two is that VSync is coincident with the start of the line, establishing the frame as field 1. This sequence is easier to generate than the official field 1 sequence as it contains no partial lines, and is often used by many video games consoles and home computers.

The ZX Spectrum departs from the specification in the interest of simplicity, and performs a cheat in which it generates one long 256μs VSync pulse that spans four scan lines, as shown in Figure 11-4 sequence D. This is enough to reset the television saw-tooth oscillator and maintain sync lock. Table 11-2 shows that this pulse sequence begins at the start of an eight line sync period.

This simple vertical synchronization component of the ZX Spectrum video signal exploits the analogue nature of the vertical control oscillator in contemporary televisions. However, some modern digital and LCD televisions have problems locking onto this vertical sync, as they process the signal digitally and do not recognise the fake VSync pulse.

As with the horizontal sync, the video output must be blanked during the vertical sync - normally for a period of eight lines. In the ZX Spectrum this blank occurs only while the 256μs VSync pulse is active, given by the following equation for the PAL version of the ULA:

$$VSync_{(PAL)} = \overline{\overline{V7} + \overline{V6} + \overline{V5} + \overline{V4} + \overline{V3} + V2}$$

V2 ensures that the vertical sync lasts for no more than four scan lines, and there is evidence that this was not part of the original design prototype, which may have had an eight line vertical sync. See the section called *The ZX Spectrum Maskable Interrupts* in Chapter 21, *Interrupts* for further information.

Note that because the vertical counter is clocked by CLKHC6, the counter does not increment with the start of each scan line, as defined by HSync, but 6.86μs later when CLKHC6 goes low. This leads to a misalignment of the vertical sync with respect to the HSync, following $6.86\mu s$ later, as shown in Figure 11-4.

Figure 11-5 shows the ZX Spectrum ULA implementation of the Border, Blank and Sync. Border is used internally to switch the video generation between pixel and border output, VSync and HSync are used by the video output multiplexer to switch off RGB colour generation (Figure 12-10), Blank and combined Sync are passed to the YUV colour difference encoding circuit, where they are combined with the colour information and passed out of the ULA to the composite video encoder.

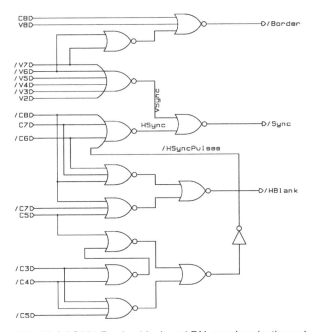

*Figure 11-5: ULA 6C001 Border, blank and PAL synchronisation schematic*

## NTSC Vertical Synchronisation

The NTSC version of the ZX Spectrum has a slightly shorter vertical display period, containing just 24 lines for the bottom border and 40 lines for the top. This means that the vertical synchronisation pulse occurs earlier than on a PAL machine, at 216 scan lines from the start of the pixel display area. See Table 11-3.

The engineers at Sinclair inverted V5 in the generation of the NTSC VSync to bring the synchronisation pulse forward by 32 scan lines. This reduced the number of lines in both the top and bottom borders, adjusting the vertical position of pixel display rectangle towards the centre of the shorter screen.

Signal V5 is routed within one matrix cell of the VSync NOR gate, making this a simple design modification that did not require the re-routing of any other interconnect tracks. The NTSC VSync is defined as:

$$VSync_{(NTSC)} = \overline{V7} + \overline{V6} + V5 + \overline{V4} + \overline{V3} + V2$$

The NTSC VSync schematic is given in Figure 11-6 and should be compared to the PAL schematic of Figure 11-5.

| Block Description | Lines | Length | V8–0 at Block Start |
|---|---|---|---|
| Display | 000 – 191 | 192 | 0 00 000 000 |
| Bottom Border | 192 – 215 | 24 | 0 11 000 000 |
| Sync Period | 216 – 223 | 8 | 0 11 011 000 |
| Sync Pulse | 216 – 219 | 4 | 0 11 011 000 |
| Top Border | 224 – 263 | 40 | 0 11 100 000 |
| Clock Reset | 264 – 264 | 0 | 1 00 001 000 |

*Table 11-3: NTSC vertical counter states and associated display regions*

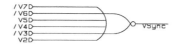

*Figure 11-6: ULA 6C011 vertical synchronisation schematic*

# Chapter 12
# Generating The Display

The ZX Spectrum display has a resolution of 256 × 192 pixels, organised as a virtual 32 × 24 character grid in which a character is one byte wide and eight bytes high. This geometry is identical to the ZX81, except that direct access to video memory is allowed where the ZX81 used a character ROM[1], and each character cell may be assigned a background and foreground colour where the ZX81 was black and white.

The colour information follows the display bytes in memory and divides the display into a grid of 32 × 24 colour cells. This colour overlay is applied to the black and white pixel display such that a cell defines a single background and foreground colour for the eight display bytes within that cell position. When working at a character level this colour layout is simple and flexible, however when working at a pixel level the two colours per character cell restriction becomes apparent, an effect known as attribute-clash[2].

Some viewed this restriction negatively, seeing it as a design flaw and an oversight. However that could not have been further from the truth. In April 1982, Richard Altwasser and Sinclair Research filed an international patent [ALTWASSERDC] citing the reduction in both memory and complexity of the Sinclair colour display, when compared to conventional displays, as the object of invention. It also states as an advantage that the colour information is at a lower resolution than would usually be the case, demonstrating that the attribute-clash effect was intentional. For some this clash may have detracted from the visual ability of the ZX Spectrum, but the memory that was saved gave programmers more freedom than with competitor machines, which paid dividends when it came to writing high quality software and games. The best programmers became apt at concealing the colour limitation by clever graphic sizing[3] or by reducing the colour content of the screen[4].

To keep the display updated the ZX Spectrum repeatedly fetches pairs of bytes from memory, one containing pixel information, the other colour. It feeds the pixel display byte through a shift register to extract each individual pixel in turn, and uses them to select whether a foreground or background colour is sent to the television. The foreground and background colours themselves are determined by the pixel's associated colour byte.

## Pixel Display Generation

There are three sources of data required to generate the ZX Spectrum's display, depending on where the electron beam is at a given time: Border colour, display byte containing pixel information and its colour attribute byte. Figure 12-1 shows the relationship between byte, pixel and character cell, and it should be recalled that a single attribute byte provides the foreground and background colour for all display bytes within a character cell.

Each bit of a display byte represents one pixel, with bit seven being the leftmost pixel of the eight. When a display byte is to be output to the screen, it is shifted one bit at a time into the colour circuit by the 7MHz pixel clock.

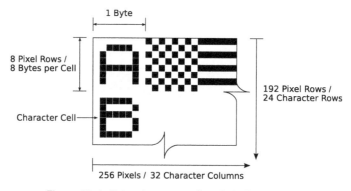

*Figure 12-1: Byte, character cell and pixel relationship*

Once eight bits have been shifted to the display, the next byte needs to be made available. At 142ns, the time between each pixel shift is insufficient to allow this new byte to be loaded from memory on demand [DS4116]. To resolve this, the ZX Spectrum employs a double byte buffer in the load and shift circuit.

The required display byte is first fetched from the comparatively slow memory into a temporary eight bit latch, from where it is loaded into the eight bit shift register when required. This theoretically allows up to $1.14\mu s$ for the ULA to fetch a display byte from memory while the shift register is serialising the current display byte, so that the next display byte is available when the shift register needs it.

Figure 12-2 illustrates the double buffer mechanism where the *Data Latch* signal loads a byte from memory via the data bus D7–0 into the eight bit transparent latch, signal *Data Load* synchronously transfers the byte from the latch into the shift register at the next negative edge of the pixel clock, and signal *Pixel Clock* shifts the byte in the register out through *Serial Out*, bit seven first.

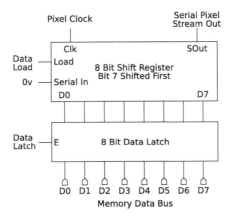

*Figure 12-2: Pixel data latch and shift register block diagram*

The serial pixel stream generated by the shift register is not fed directly to the display, as this would produce a black and white image, but instead switches a colour generation circuit between a foreground and background colour specified by the attribute byte associated with the pixel's screen position, the choice depending on whether the pixel is set or not.

| Modifier | | Background | | | Foreground | | |
|---|---|---|---|---|---|---|---|
| 7 | 6 | 5 | 4 | 3 | 2 | 1 | 0 |
| FL | HL | G | R | B | G | R | B |

*Figure 12-3: Attribute byte bit assignments*

The ZX Spectrum has a three bit RGB colour palette, so six bits of the attribute byte are require to represent both the foreground and background colours. The remaining two bits are designated colour modifiers, such that bit six enables a highlight or brightness increase for the attribute, and bit seven enables a flash mode for the attribute, where the foreground and background colours are cyclically swapped every half second or so.

The lower six bits of the attribute byte, giving the foreground and background colours, are passed to a three channel, 2-to-1 multiplexer which is switched by the serial pixel stream. When the serial pixel output is high, indicating that a pixel is set, the foreground colour is selected (bits D2–0 through channel A in Figure 12-4), otherwise it selects the background colour (D5–3 through channel B). The selected RGB colour is then passed to the analogue video generation circuit, discussed in Chapter 16, *Analogue Video*.

Each time a display data byte is loaded into the shift register, the corresponding attribute byte is presented to the colour output multiplexer, where it remains while the eight pixels are shifted to the display. The timing of the attribute byte presentation is critical. The pixel shift register loads data at the next negative edge of the pixel clock, and the attribute byte must appear at the multiplexer at exactly the same time. Too early and the last pixel of the preceding display byte will change to the new colours before it has finished being displayed, too late and the new pixel will begin to be displayed with the colour of the previous attribute.

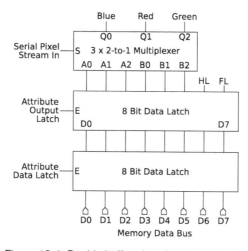

*Figure 12-4: Double buffered attribute byte fetch block diagram*

To ensure that the new attribute byte appears at the multiplexer at the same time as the first shifted bit of the new display byte, the ZX Spectrum feeds the colour output multiplexer from an eight bit attribute output latch that is loaded by a signal that is synchronous with the loading of the shift register. Figure 12-5 illustrates this timing and shows a new data byte being loaded into the shift register every eight pixel clock cycles by the combination of the SLoad signal and negative clock edge, while simultaneously loading a new attribute byte into the attribute output latch with the /AOLatch signal.

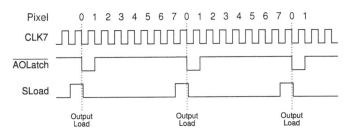

*Figure 12-5: Attribute output latch, SLoad and clock relationship*

Fetching the attribute byte from memory directly into the output latch is not possible due to the demanding timing requirements discussed previously, so Altwasser employed the same double buffering technique used with the data byte, shown in Figure 12-4. An attribute byte is first fetched from memory into an intermediate attribute data latch, from where the attribute output latch is loaded as described above. This relaxes the constraints on when the data and attribute bytes must be fetched from memory, so long as the two memory data latches are loaded before the bytes are required for output, which is a 1.14s window (the time taken to shift out eight pixels).

## The Flash Mode

When the flash mode bit of the attribute byte is set, the ULA swaps the foreground and background colours that are sent to the screen for that attribute byte at approximately half second intervals. See Chapter 14, *Video Control Clocks* for further details of the flash rate.

The ULA does this, not by physically exchanging RGB values, but by inverting the pixel stream that is sent to the colour output multiplexer. By taking the flash enable attribute bit and NOR gating this with the flash clock, the ULA creates a signal that is low when the flash enable bit is reset or oscillating at the flash clock rate when the enable is set. The serial pixel stream is XNOR gated with this flash control signal, causing it to invert and revert with each cycle of the flash clock when the flash mode is enabled.

The flash modification of the serial pixel stream output by the shift register is illustrated by the following equation:

$$\overline{SerialPixelStream} = ShiftRegOut \oplus \overline{(FlashMode + FlashClock)}$$

## Border Generation

When the ZX Spectrum is not generating the pixel display area, the coloured border around it is produced, see Figure 9-1. During this period no data is fetched from the memory and no pixels are shifted out of the shift register; therefore the last background colour held by the attribute output latch would be sent to the screen.

To allow the border to be set to a specific colour, the ULA provides an additional three bit latch to hold the RGB value for the border colour. The value stored by the latch is set by software writing to an input port assigned to the ULA; see Chapter 19, *Input-Output Devices*.

As the ZX Spectrum video controller falls back to producing a background colour in the absence of a serial pixel stream, the RGB value provided by the border colour register needs to be selected instead, whenever the pixel display area is not being generated.

The VidEN signal, which is active while the pixel display area is being produced, switches a five channel 2-to-1 multiplexer to select between the background colour and border colour. Note that both highlight and flash are disabled during the border region, which accounts for channels four and five in addition to the three RGB channels. Theoretically, there are two places where this border colour multiplexing could occur in Figure 12-4; either between the attribute output latch and multiplexer or between the attribute data latch and output latch.

If the border/background colour multiplexer were placed between the attribute output latch and output multiplexer, additional propagation delay would be introduced into the background channel, which would not be present in the foreground channel. This asymmetric delay would cause the background colour of each screen attribute cell to bleed into the cell to its right, as the transition to the new background colour would be slower than to the new foreground colour.

Therefore Altwasser placed the border/background colour multiplexer between the data and output latches, so that the output latch hides the additional propagation delay. See Figure 12-6. This is the more complicated of the two placements, as the output latch now needs to be clocked whenever a switch from the pixel display to border is made, and whenever the border colour is changed. For this reason the ULA attribute output latch signal contains a continuous chain of pulses, one every eight cycles of the 7MHz clock. See Chapter 14, *Video Control Clocks* and Figure 12-5.

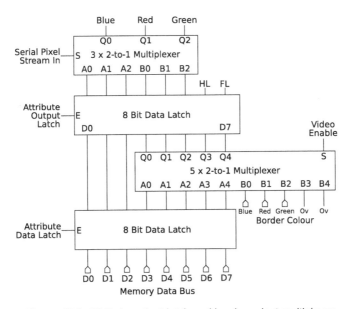

*Figure 12-6: Attribute output latch and border select multiplexer*

The consequence of this multiplexer placement is that it is not possible to change the border colour more frequently than once every eight pixel clock cycles. Carefully written code can choose where the electron beam will be when it sets the border colour register, but the border itself will not change until the attribute output latch is triggered.

## Control Signals

The timing of the byte fetches from memory must be carefully chosen to fall within the specifications laid out by the dynamic RAM data sheet [DS4116]. There are minimum periods that the memory control signals RAS and CAS must be active before the requested byte on the data bus is stable enough to be read; therefore the set up time for each byte read is expensive.

The time between the display and attribute byte fetches is minimised by utilising a *page mode* read [DS4116], where a row address is presented once to the memory, followed by two successive column addresses. This results in two bytes being read from the same row in the memory and avoids the set up time of supplying a second row address. To enable this type of read, the memory structure of the ZX Spectrum is carefully arranged so that an attribute byte is in the same memory row as its display byte, leading to the peculiar screen byte order and the characteristic way that screen pictures appear on the television when loading software from cassette. See Chapter 15, *Video Addressing* for further details of this memory layout.

As the display controller shares the video memory with the CPU, there is a small set up delay while it assumes control. To make access more efficient the controller reads two complete pairs of bytes each time it has control; therefore halving the number of memory claims and reducing the set up delays necessary.

To perform this operation, once the display controller has loaded a pair of bytes into the memory latches and transferred them to the shift register and attribute output latch, it immediately loads the next pair into the now empty memory latches, making effective use of the limited time window available.

Before the specifics of video control signal generation can be discussed, an understanding of the video memory access timing implemented by the ZX Spectrum is required, which is presented in Chapter 13, *Video Memory Access*. Chapter 14, *Video Control Clocks* then describes the video control signal generation in detail.

At this point the control signals may be usefully summarised as:

1. *CLK7*. The 7MHz pixel clock that controls when pixels are output to the display by driving the shift register.

2. *VidEN*. This active high signal indicates when the pixel display is being generated, otherwise the border is generated.

3. *DataLatch*. The active low signal that transfers the byte on the ULA data bus into the display data latch.

4. *AttrLatch*. The active low signal that transfers the byte on the ULA data bus into the attribute data latch.

5. *SLoad*. The active high signal that instructs the shift register to load the display byte from the data latch at the next negative edge of the pixel clock.

6. *AOLatch*. The active low, pixel clock synchronous signal that transfers an attribute byte from the attribute data latch (or border colour register) into the attribute output latch. This signal goes low every eight cycles of the 7MHz pixel clock.

These control signals activate in a carefully choreographed sequence that fetches and passes display and attribute bytes through the display circuitry. How this timing is achieved is described in Chapter 14, *Video Control Clocks*, and results in the following sequence of events:

1. DataLatch reads the display byte on the memory data bus and stores it in the display data latch.

2. AttrLatch reads the attribute byte on the memory data bus and stores it in the attribute data latch.

3. The shift register clocks the last pixel of the previous display byte through its serial output. Following this, SLoad goes high ready to reload the shift register from the display data latch at the next pixel clock.

4. With the next transition of the pixel clock the shift register loads display data from the data latch and AOLatch goes low, reloading the attribute output latch which feeds the colour output multiplexer.

5. Once the clock transition is complete and the shift register has been loaded, SLoad returns low and AOLatch high. DataLatch then reads and stores a second display byte from the data bus into the display data latch.

6. AttrLatch reads and stores a second attribute byte from the data bus into the attribute data latch.

7. After several more transitions of the pixel clock, when the last pixel of the current data byte has been clocked through the shift register output, SLoad goes high to reload the shift register from the preloaded display data latch, at the next downwards transition of the pixel clock.

8. The next transition of the pixel clock reloads the shift register while AOLatch goes low, transferring the next attribute byte into the output latch which feeds the colour output multiplexer. When the downward transition is complete SLoad returns low, followed a little later by AOLatch going high.

The sequence restarts a little before the shift register sends the last pixel of the display byte to the television, so that the reload of the shift register (3) occurs as soon as it is empty.

## Circuit Operation

### Display Byte Latch and Shift Register

At its most fundamental level, the ZX Spectrum ULA updates the screen by serialising bytes representing the on/off states of pixels, where one bit represents one pixel, into a stream of bits which are encoded and sent to the television. Figure 12-7 shows the data latch and shift circuit that performs this serialisation.

The data latch consists of eight gated D transparent latches, sharing a common latch enable. When the enable is pulled low by /DataLatch, the latch becomes transparent and the value on the data bus is seen at the latch output Q7–0. When /DataLatch returns high, the value at the latch output is stored. Display bytes are loaded into the data latch by /DataLatch, the exact timing of which is discussed in Chapter 14, *Video Control Clocks*, suffice it to say that a display byte is loaded into the latch ahead of being required by the shift register.

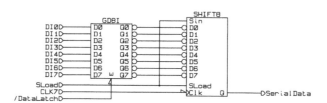

*Figure 12-7: Data latch and shift register schematic*

The display data latch feeds an eight bit shift register (described in Appendix B, *Component Library*), which clocks bits out of its serial output on the downward transition of the 7MHz pixel clock. After eight bits, and therefore pixels, have been shifted out, the shift register will be empty and require reloading.

The SLoad signal therefore goes high for the first of every eight downward pixel clock transitions, forcing the shift register to load a new byte from the display data latch and transfer the first bit to its output.

Note that the schematic in Figure 12-7 shows an inverted value being passed from the latch to the shift register. For efficiency the ULA implementation of the shift register inverts the value passed to it; therefore the register requires an inverted value to be supplied to correct this behaviour. Since the transparent latches provide both inverted and non-inverted outputs, passing an inverted value to the shift register requires no additional logic.

**Attribute Data Latch and Border Multiplexer**

Like the display data latch, the attribute data latch also consists of eight gated D transparent latches with a common enable. As /AttrLatch goes low, the attribute value on the data bus is transferred to the latch output and is stored when /AttrLatch returns high. Both the latch signal and memory read are carefully timed so that the attribute byte is latched and made available to the output latch before being required.

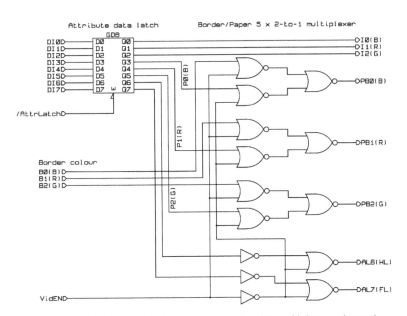

*Figure 12-8: Attribute data latch and paper/border multiplexer schematic*

Bits Q7–3 of the latched byte, containing the background RGB value and attribute modifier bits, are fed to the output latch via five 2-to-1 multiplexers, where they are combined with the RGB value of the border colour taken from the border colour register. The multiplexer channel select is controlled by the VidEN signal, which is high during pixel stream output and low during the border; therefore replacing the background colour with that of the border whenever the pixel display area is not being produced.

The RGB channels of the multiplexer consist of three NOR gates, two of which gate the input channels with the channel select, and a third which combines the result of the channel selection into a single output, as shown in Figure 12-8 The remaining two channels are the flash and highlight channels which have a single input source fed from the attribute data latch. While the border is being produced, the outputs of these two multiplexers are explicitly forced to zero, disabling the highlight and flash mode.

### Flash Control of Pixel Stream

The flash control circuit inverts the flash mode select bit and NOR gates this with the flash clock, so that when the flash mode bit is reset, the flash clock is blocked and the output of the NOR is forced low. When the flash mode bit is set, the flash clock is passed through the NOR gate where it is inverted. See Chapter 14, *Video Control Clocks* for details of the flash clock generation. The flash mode select is taken from the attribute output latch described in the section called *Attribute Output Latch and Output Multiplexer*, because this is synchronised with the change in attribute colour for the current display byte, and the switch to border colour generation.

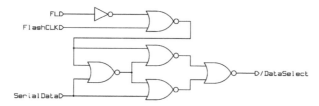

*Figure 12-9: Flash control of pixel stream schematic*

This flash control signal is combined with the serial pixel stream through a classic XNOR gate constructed from four NOR gates, so that while the flash mode bit is set — which passes the flash clock to the XNOR gate — the serial pixel stream is alternately inverted and reverted. This provides the cyclic pixel invert necessary to flash the associated screen attribute cell.

### Attribute Output Latch and Output Multiplexer

The colour output stage show in Figure 12-10 combines the serial pixel stream from the flash control circuit with the output of the attribute data latch and border colour multiplexer, creating an RGB output that produces the foreground or background colour, depending on the state of the stream.

The colour and modifier bits from the attribute latch and paper/border colour multiplexer are fed to a gated D transparent latch, controlled by /AOLatch which goes low precisely as the shift register is loaded and the first bit of the display byte appears in the serial pixel stream.

The colours stored by the latch are multiplexed by a three channel, 2-to-1 multiplexer, which switches between the foreground or background/border colour depending on the state of the serial pixel stream, /DataSelect.

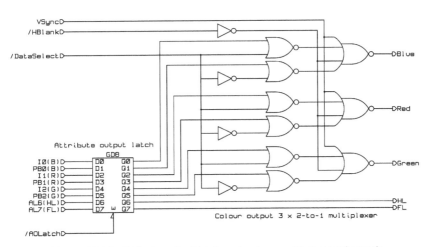

*Figure 12-10: Attribute output latch and colour multiplexer schematic*

Each of the three output multiplexer channels that together produce the final RGB signal, consist of three NOR gates, two of which gate the foreground and background/border colour with the /DataSelect signal, and a third which combines the gate outputs with the vertical synchronisation and horizontal blanking signals. Whenever these synchronisation or blank signals are active, the multiplexer channel output is forced low, disabling or blanking the colour output during periods of synchronisation.

The flash modifier bit from the output multiplexer is fed back to the flash control circuit of Figure 12-9.

Finally, the RGB signals are fed to the analogue video circuit, discussed in Chapter 16, *Analogue Video*, along with the highlight modifier bit.

---

1.  Because the ZX81 display is updated by software and the Z80 processor, programmers eventually discovered how to take control of this process to produce high resolution graphics. This technique consumes nearly all of the CPU's resources however.

2.  The colours, brightness and flash at a given character position are called attributes.

3.  Exolon, developed by Raffaele Cecco, published by Hewson Consultants

4.  Knightlore, developed and published by A.C.G under their Ultimate Play The Game label.

# Chapter 13
# Video Memory Access

The ULA generates control signals for the lower 16K dynamic RAM for itself and on behalf of the CPU. As access to this memory is shared, the ULA and CPU control signals must be combined. This is performed within the ULA to allow it to detect when the CPU is about to access the memory, reduce duplication and simplify the circuit external to the ULA.

The operation of dynamic RAM is described in the section called *Dynamic RAM* in Chapter 3, *The Standard Microcomputer*.

The CPU address bus is multiplexed into two seven bit row and column addresses by two external multiplexer chips, and fed to the DRAM via seven in-line resistors. The resistors provide isolation between the CPU and ULA address buses, the ULA being directly connected to the DRAM, so that when the ULA presents an address to the memory, it overrides any CPU address that may be on the bus. This is discussed further in the section called *16K DRAM CPU Interface* in Chapter 17, *CPU Memory Access*.

## Video Access Control

The DRAM control signal timing for video access is closely related to the video fetch timing discussed in Chapter 12, *Generating The Display*. Remember that bytes are fetched from the RAM in pairs, with two pairs being read in quick succession via a page mode read. The timing of the byte fetches are carefully related to the pixel stream being sent to the television and the internal buffer handling, so that each pixel and attribute byte pair is ready for use before the display controller has finished with the current pair, and buffers are reloaded as soon as they have been emptied.

In addition to the row and column address strobe signals, the DRAM also requires a row and column address to be placed on its address bus. As the ULA is connected directly to the DRAM, the ULA keeps its address bus in a high impedance state when not performing a video fetch. When a fetch is to be carried out, its address bus is enabled by an internal control signal and the CPU address on the bus is overridden.

When the CPU requests access to the lower 16K of memory during a video memory fetch, the ULA stops sending the CPU its clock signal; therefore halting time for the CPU. The video fetch continues until two pairs of bytes have been read, at which time the video controller disables its address bus and re-enables the CPU clock. The CPU will not be aware that time had stopped, and continues with the read or write operation it had begun. See Chapter 18, *CPU Clock and Contention* for details on the handling of memory contention.

The generation of appropriate row and column addresses for each RAS and CAS pulse, when the video address bus is enabled, is discussed in Chapter 15, *Video Addressing*.

## Page Mode Read

Page mode allows multiple column addresses to be read from or written to in a single address row. The row address strobe, RAS, is held while strobing new column addresses with CAS. This avoids the set up and hold times associated with the row address and allows faster access to multiple locations.

Figure 13-1 and Table 13-1 give the page mode read timings specified by the $\mu$PD416 datasheet for the 150ns device used in the ZX Spectrum [DS4116]. All times are given in nanoseconds.

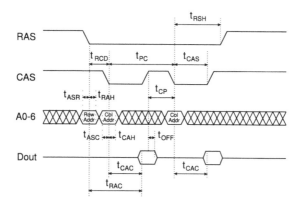

*Figure 13-1: 4116 dynamic RAM page mode read signal waveforms*

The most important information to derive from the table is that there must be 20ns between RAS and the first CAS, and 60ns between each CAS. RAS must be held for a minimum of 100ns from the start of the last CAS, and data is available on the data bus after 100ns from the CAS start, until CAS is removed.

During a read operation, the DRAM write enable line, /WE, must be high for the duration of the CAS pulse.

| Description | Parameter | Min | Max |
|---|---|---|---|
| Row address set-up time | $t_{ASR}$ | 0 | |
| Column address set-up time | $t_{ASC}$ | -10 | |
| Row address hold time | $t_{RAH}$ | 20 | |
| RAS to CAS delay time | $t_{RCD}$ | 20 | 50 |
| Page mode cycle time | $t_{PC}$ | 170 | |
| RAS precharge time | $t_{RP}$ | 100 | |
| CAS precharge time | $t_{CP}$ | 60 | |
| CAS pulse width | $t_{CAS}$ | 100 | 10000 |
| RAS hold time | $t_{RSH}$ | 100 | |
| Column address hold time | $t_{CAH}$ | 45 | |
| Access time from RAS | $t_{RAC}$ | | 150 |
| Access time from CAS | $t_{CAC}$ | | 100 |
| Output buffer turn-off delay | $t_{OFF}$ | 0 | 40 |

*Table 13-1: 4116-150ns dynamic RAM page mode read timing (ns)*

## RAS and CAS Timing Overview

The timing of the RAS and CAS signals, and the related internal latch signals DataLatch and AttrLatch, originate from the need to complete the byte fetches as quickly as possible to limit the impact on the CPU. All signals are derived from the 7MHz clock and master counter and therefore have a resolution of 72ns, which is half the 7MHz clock period. That said, some signals are deliberately subjected to propagation delay to tune their timing, usually in the order of tens of nanoseconds. Figure 13-2 shows the timing of the RAS and CAS pulses generated by the video controller, and their relationship to the associated byte latch signals (which are described fully in Chapter 14, *Video Control Clocks*.

When the video controller reads a byte pair, it generates a single RAS pulse along with two CAS pulses. The length of the RAS pulse depends on the RAM specification and the length of each CAS pulse, as RAS must be active when CAS is applied. The datasheet states that a CAS pulse must be at least 100ns in duration, and that data will be available no sooner than 100ns after its start. The video controller must therefore wait 100ns after it issues a CAS before it can read the data bus, and cannot remove the CAS until the fetch is complete ($t_{OFF}$). Re-factoring these figures into multiples of 72ns gives a theoretical minimum CAS duration of 144ns which, with a data bus fetch and latch period of 72ns, gives an measured CAS width of approximately 216ns.

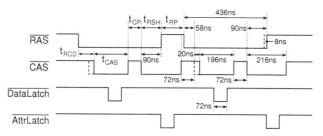

*Figure 13-2: 6C001E-7 video RAS, CAS and byte latch timing*

The second CAS cannot occur for another 60ns, as specified by $t_{CP}$, so the video controller waits for 72ns before it generates the next 216ns CAS pulse.

The RAS signal must be applied $t_{RCD}$ (20ns) before the first CAS, and Alt-wasser created an appropriate $t_{RCD}$ by combining two techniques; One, generating RAS and CAS at the same time and chopping off approximately 20ns from the start of the first and third CAS pulses with some intentionally delayed logic (dashed section at the start of CAS pulse in Figure 13-2). Two, by delaying the CAS signal with some propagation delay. This creates a total RAS to CAS delay of approximately 78ns, in the case of the 6C001E-7 ULA.

The second pair of byte fetches may follow the first after the minimum RAS to RAS delay of $t_{RP}$ (100ns). If RAS were to end at the same time as the second CAS pulse, then the ULA would have to delay the third CAS pulse by $t_{RP}$ + $t_{RCD}$ (RAS to RAS delay + RAS to CAS delay), which equates to 120ns. To complete the four byte read in as short a time as possible, Altwasser required a maximum period of 72ns to separate each CAS pulse, and achieved this by exploiting a property of the RAS that allows it to be removed before the end of the CAS, so long as it has been held for $t_{RSH}$ns from the CAS start.

At 216ns, the second CAS pulse is over twice as long as the specified $t_{RSH}$ of 100ns, allowing RAS to be safely removed 72ns before the CAS ends.

Figure 13-3 shows a capture of the second RAS with the third and fourth CAS pulses for a 6C001-E ULA, showing the RAS pulse to be 435.6ns in duration. Grid divisions are at 40ns.

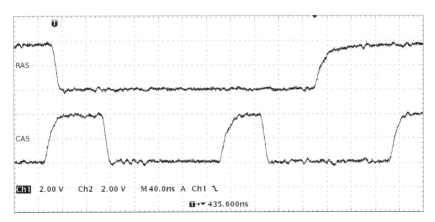

*Figure 13-3: 6C001E-7 video RAS and CAS signal capture*

## RAS Generation

The generation of the RAS signal is carried out by first creating a chain of pulses of a suitable length and spacing, and then selecting pulses from this chain as required.

As discussed previously, a RAS pulse spanning the duration of both CAS pulses will be active for 504ns. This RAS pulse is shortened by 72ns to allow the third CAS to go low 72ns after the second, giving a RAS duration of 432ns.

Referring to Figure 13-4, Altwasser created a chain of RAS pulses of this duration by taking the duration of one half period of C1 and one half period of C0, defining VidRASPulse as:

$$\overline{VidRASPulse} = \overline{\overline{C0} + \overline{C1}}$$

123

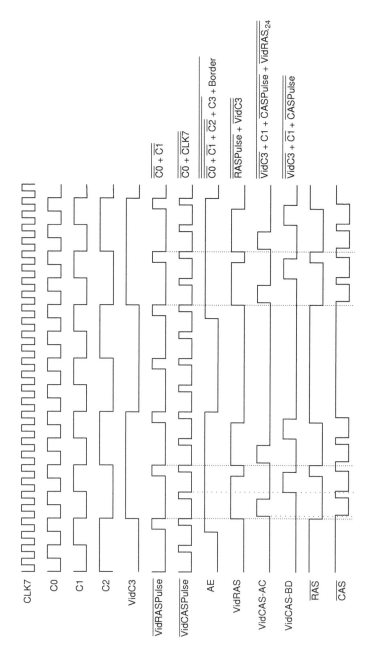

*Figure 13-4: Video RAS and CAS generation*

The required video RAS pulse pairs are selected from the VidRASPulse chain by VidC3, a copy of C3 that is active only when generating the pixel display rectangle and described in the section called *DataLatch* in Chapter 14, *Video Control Clocks*. The VidRAS signal is therefore defined as:

$$VidRAS = \overline{\overline{VidRASPulse} + \overline{VidC3}}$$

Both the internal VidRAS and final output /RAS signals are shown in Figure 13-4, and should be compared to Figure 13-2.

## CAS Generation

Video CAS generation is similar to that of RAS, in that a chain of CAS pulses of suitable length and spacing is created, from which appropriate pulses are selected.

As show previously, for the given clock and DRAM speed, the desired CAS duration is 216ns. Referring to Figure 13-4, this represents one and a half periods of CLK7 or, in terms of C0 and CLK7, the VidCASPulse chain is defined as:

$$\overline{VidCASPulse} = \overline{\overline{C0} + \overline{CLK7}}$$

Two CAS pulses are selected from this chain for each RAS pulse. The first occurs at the beginning of the RAS pulse to fetch the display byte from the memory row, the second at the end of the RAS pulse to fetch the attribute byte. See the page mode read timing in Figure 13-1.

To introduce the necessary page mode RAS to CAS delay of $t_{RCD}$, the first CAS pulse of the pair needs to be shorter than the second, and start later; therefore the pulses are shorter than the second, and start later; therefore the pulses are selected separately from the chain, modified and then combined into a single signal.

As a display fetch cycle consists of two RAS pulses, two pairs of CAS pulses are required, the first and third (A and C) CAS pulses being selected by:

$$VidCAS_{AC} = \overline{\overline{VidC3} + C1 + \overline{VidCASPulse} + \overline{VidRAS_{-delay}}}$$

By using a delayed VidRAS, the beginning of the selected CAS pulse is chopped off, holding its start until approximately 20ns (6C001E-7) after RAS has gone low. This ensures that RAS is activated first, and along with the later delay of VidCAS, produces a RAS to CAS delay of 78ns, complying with the specification $t_{RCD}$ of 20ns.

The second and fourth (B and D) CAS pulses are selected by:

$$VidCAS_{BD} = \overline{\overline{VidC3} + \overline{C1} + \overline{VidCASPulse}}$$

The combined video CAS signal is generated by NORing these two component CAS pulse signals together:

$$\overline{VidCAS} = \overline{VidCAS_{AC} + VidCAS_{BD}}$$

RAS and CAS pulse timings for a 6C001E-7 ULA are given in Figure 13-2.

## Circuit Description

The video RAS and CAS generation circuit is shown in Figure 13-5. It is split into several sections: RAS and CAS pulse generation, RAS and CAS pulse selection and the merging of these RAS and CAS signals with their CPU equivalents into combined /RAS and /CAS outputs.

At several points in the circuit inverters have been used to provide some propagation delay or signal buffering. This was done to align the timing of signals to avoid glitches, delay signals with respect to one another or to counteract the effects of interconnect track length and parasitic capacitance on signal transition times and delays.

Some of this buffering appears unnecessary, even after considering gate fanout and track length. For instance, the generation of /VidRASPulse NOR gates /C0 with /C1 *and* a delayed /C0, having been passed through four inverters.

Altwasser says that he added buffers and re-routed interconnect tracks to address signal delay and slow rise/fall times due to track length and parasitic capacitance, particularly in the area of memory timing (see Chapter 6, *Sinclair and the ULA*). However, it is not clear why combining a delayed /C0 along with /C0 in the /VidRASPulse generation would have been necessary, and only appears to lengthen the RAS pulse by 8ns.

The selection of CAS pulses for the /VidCAS$_{BD}$ signal uses a delayed /C1, whereas its companion signal, /VidCAS$_{AC}$, uses a non-delayed C1. As both use the same source C1 signal, the delay does not appear to serve any obvious purpose. The delay of /C1 in /VidCAS$_{BD}$ could have been included to mirror that of /VidRAS in /VidCAS$_{AC}$, but their effects are very different in each case.

Other source and generated signals are delayed to provide a functional benefit. For instance, the generation of the /VidCASPulse chain uses six inverters to delay /CLK7 and align with /C0, since /C0 will have undergone some delay in its production, with respect to CLK7.

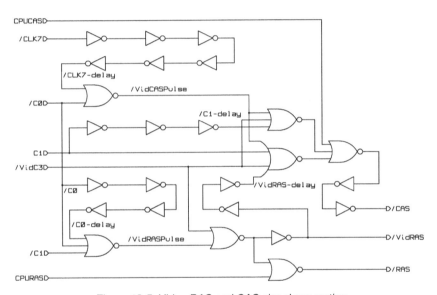

*Figure 13-5: Video RAS and CAS signal generation*

The propagation delay incurred by the generation of VidCAS$_{AC}$ and VidCAS$_{BD}$ is increased by two additional inverters, creating a /CAS signal that is delayed with respect to /RAS by approximately 58ns, as measured on a 6C001E-7 ULA. This ensures that the data and attribute byte fetches complete before /CAS is removed, and with the 20ns chopped from the front of the first and third CAS pulses, increases $t_{RCD}$ to 78ns.

The consequence of delaying /CAS is that the time between the CAS start and RAS end ($t_{RSH}$) is shortened, which in the case of the 6C001E-7 ULA with its measured CAS delay of 58ns, reduces $t_{RSH}$ from 148ns to 90ns, and does not meet the DRAM specification of 100ns. Other versions of the ULA have a different CAS delay factor, for example the 6C001E-6 ULA has a measured delay of approximately 20ns, resulting in a $t_{RSH}$ of 128ns. The 6C001E-7 is the most common ULA use in the ZX Spectrum, and its out of specification timing does not cause any detrimental effects.

Chapter 22, *Signal Interfacing* describes the peripheral cell interface of the ULA RAS and CAS signals out to the 4116 16K DRAM, but /RAS deserves special mention here as its output is provided by a tri-state totem pole interface that should be enabled whenever the ULA video generator or CPU require access to the 16K DRAM. However, due to a design error, the /RAS output is almost always enabled, even when the DRAM is not being accessed. Consequently, this enable signal is not discussed here, and it is given full consideration in the section called *DRAM Row Address Strobe* in Chapter 22, *Signal Interfacing* and the section called *Disabled 16K DRAM Refresh* in Chapter 23, *Hidden Features and Errors*.

# Chapter 14
# Video Control Clocks

Now that the video byte fetch timings have been defined and understood, the control signals that read the data bus and process the information through the video controller can be discussed.

The control signals shown in Figure 14-1 illustrate the loading of display and attribute bytes from the video memory, and the timing of their transfer through the video circuit. The control signals may be divided into two groups: those which transfer data from the data bus into memory latches, and those which transfer data from the latches into the video output circuit.

The signals are closely related to the video memory control signals, in particular CAS, so that a byte on the data bus is fetched into the relevant memory latch as soon as it is available. The close signal relationship allows the latched bytes to be transferred into the video output circuit as soon as they have been loaded, freeing the latches for further bytes, loaded while the video generator still has control of the memory.

## Sequence Overview

The video control signals are choreographed in such a way as to produce the following sequence of events:

1. The DataLatch pulse labelled *D1* stores the first display byte into the display memory latch, followed by AttrLatch pulse *A1*, which stores the first attribute byte into its memory latch.

2. The two latched bytes are transferred into the shift register and attribute output latch respectively by *AOLatch* and *SLoad*, which are aligned with the pixel clock at time-point *Output Load*.

3. Immediately following this, a second pair of bytes are loaded and stored in the now empty memory latches by *D2* and *A2*. Here they and held until

the second *Output Load*, eight pixel clock cycles after the first, when they too are transferred into the shift register and output latch.

It can be seen therefore, that display and attribute bytes are loaded from memory in groups, spanning the end of one output period and the beginning of the next. Once loaded, a byte pair is transferred to the output circuit at the start of the following period.

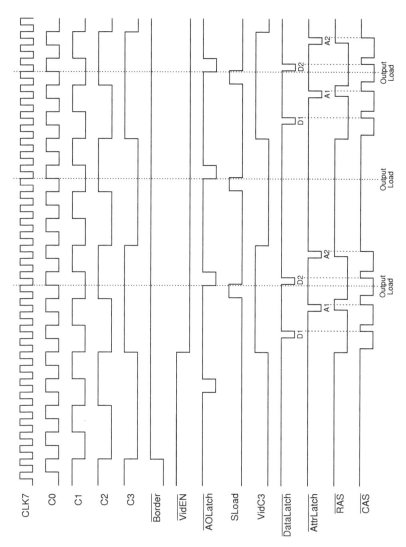

*Figure 14-1: Latch and shift register control clocks*

## Circuit Operation

Figure 14-2 shows the circuit that implements the video control signals illustrated in Figure 14-1. Its operation is straightforward even though some of the signal paths incorporate chains of inverters to introduce propagation delay and adjust their timing relative to the other signals. Each signal and its generation is discussed in detail below.

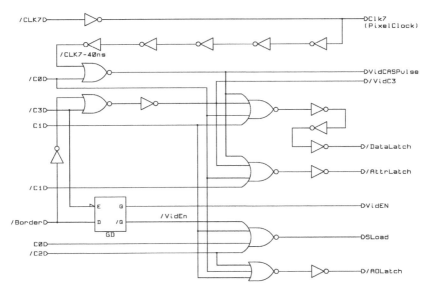

*Figure 14-2: 6C001 latch and shift register control clocks schematic*

### CLK7

The pixel clock is derived directly from the internal 7MHz clock, which is distributed in an inverted state throughout the circuit. Therefore the /CLK7 signal is inverted to produce the actual pixel clock, and for the generation of the other control signals described below it is buffered through six inverters to delay it by approximately 40ns. This is necessary to re-align the clock with the counter bits C3–0, which will themselves have undergone some propagation delay in the course of their generation. The first inverter in the chain is required to flip the clock back to a positive phase, as described in the section called *The 7MHz Clock* in Chapter 10, *The Internal Clocks*, and therefore does not contribute to the overall delay.

**VidEN**

Counter bits C8–0 give the horizontal position of the electron beam on the screen, with zero occurring at the left edge of the pixel display, such that C8 is low while the 256 pixel row is being displayed. As C8 is a component of the Border signal, it too will be low.

/Border therefore indicates that pixel generation has begun, but because no data will have been read from memory at this point, it cannot be used to activate pixel output; therefore it merely enables data loading. A second signal, VidEN is instead created which enables pixel output once a display and attribute byte pair have been fetched from memory.

This VidEN signal is generated by delaying /Border with a transparent latch until C3 has gone high, a delay of $1.14\mu s$. This is sufficient to prevent SLoad, which is controlled by /VidEN, from being activated before the first byte pair have been loaded into the memory latches (see Figure 14-1 and the section called *SLoad*).

**DataLatch**

DataLatch transfers the byte present on the ULA data bus into the eight bit transparent display memory latch. It is an active low signal that puts the latch into transparent mode until it goes high, at which point the byte is held.

DataLatch is generated as an active high signal which is later passed through three inverters to convert it into an active low signal and introduce approximately 24ns of delay. The delay ensures that CLK7 goes low first, in order to transfer the byte currently held by the latch into the shift register before DataLatch replaces it with the new byte.

DataLatch is given by:

$$\overline{DataLatch} = \overline{(\overline{CLK7 + \overline{C0}})} + \overline{C0} + C1 + \overline{VidC3}$$

Simplifying this by applying DeMorgan's Theorems gives:

$$\overline{DataLatch} = (CLK7 \cdot C0) + \overline{C0} + C1 + \overline{VidC3}$$

Where

$$\overline{VidC3} = Border + \overline{C3}$$

Thus Altwasser's implementation of DataLatch can be shown to contain the AND product of C0 and CLK7, which can be further shown to be redundant according to the distributive law:

$$(CLK7 \cdot C0) + \overline{C0} \equiv (CLK7 + \overline{C0}) \cdot (C0 + \overline{C0}) \equiv CLK7 + \overline{C0}$$

DataLatch is closely related to the dynamic memory column address strobe, CAS, which completes the memory request and causes the display byte to be placed on the data bus. A short time later, when the byte is stable enough to be read and *CAS* is being removed, DataLatch goes high, transferring the byte into the display memory latch.

The product (CLK7 • C0) is therefore, unsurprisingly, used in the generation of the CAS signal, the full equations for which are almost identical to those for DataLatch and AttrLatch. Presumably Altwasser used the same core logic for the generation of the latch and CAS signals so there would be no issue with their alignment, even though, ultimately, DataLatch is deliberately delayed by 24ns. See the section called *CAS Generation* in Chapter 13, *Video Memory Access* for further details of the CAS signal. For clarity, the VidCASPulse generation is duplicated on Figure 14-2 and Figure 13-5.

The inclusion of VidC3 restricts the generation of DataLatch pulses so that two are produced whenever C3 is high, and only within the pixel display rectangle.

**AttrLatch**

The AttrLatch signal is similar to DataLatch and transfers the byte present on the ULA data bus into the eight bit transparent attribute latch. It is an active low signal that puts the latch into transparent mode until it goes high, when the byte is held.

It is generated as an active high signal that is then inverted into an active low. Unlike DataLatch, there is no specifically introduced propagation delay.

AttrLatch is given by:

$$\overline{AttrLatch} = \overline{(\overline{CLK7} + \overline{C0})} + \overline{C0} + \overline{C1} + \overline{VidC3}$$

As with DataLatch, the redundant C0 product with CLK7 is used, borrowing the same logic found in DataLatch and CAS generation. See the section called *DataLatch* and the section called *CAS Generation* in Chapter 13, *Video Memory Access* for further details of the DataLatch and CAS signal.

As with the DataLatch signal, VidC3 is used to restrict the generation of Attr-Latch pulses to periods when C3 is high, and only while the electron beam is within the pixel display rectangle.

## SLoad

The shift register data load signal shown in Figure 12-2, referred to as SLoad in the schematics, loads the currently latched display byte into the shift register at the negative edge of the pixel clock. It is active once every eight pixel clock cycles, while /VidEN is low, reloading the shift register with a new display byte as soon as it is empty, precisely when the next pixel is required.

SLoad is given by:

$$SLoad = \overline{C0 + C1 + \overline{C2} + \overline{VidEN}}$$

Without the inclusion of /VidEN, SLoad pulses would be generated every $1.14\mu s$, and not restricted to the pixel display area. If /Border were used instead of /VidEN, the first SLoad pulse would occur $1.14\mu s$ too early. Refer to Figure 14-1, noting where the first *SLoad* pulse would occur if /Border had been used instead of /VidEN.

Because of the propagation delay inherent in each bit of the shift register, SLoad must be held high long enough through the downward transition of CLK7 for the shift register to successfully latch the new byte. As SLoad is derived from the horizontal counter, which lags behind CLK7, there is enough delay in its generation for this to be the case. See Figure 14-1.

## AOLatch

The Attribute Output Latch signal (Figure 12-4), referred to as AOLatch in the schematic of Figure 14-2, is activated repeatedly across the full width of each scan line, every eight pixel clock cycles. It is not restricted to the pixel display rectangle, as all screen colour changes are synchronised to it through the single colour output latch, including those of the border. This explains why the display output does not and cannot switch between border and pixel areas at /VidEN, but at the /AOLatch that follows it.

It is an active low signal that puts the eight bit colour output latch into transparent mode, immediately passing the attribute byte onto the output multiplexer. When it returns high at the next negative edge of CLK7, the byte is held.

Its downward transition is synchronous with the negative edge of the pixel clock CLK7, and timed so that when it occurs during a sequence of pixel output, it goes low while SLoad is low. This presents the attribute byte to the colour output multiplexer at the exact moment the display byte is loaded into the shift register, when the first pixel of which is shifted out to the multiplexer.

AOLatch is given by:

$$\overline{AOLatch} = \overline{C0} + C1 + \overline{C2}$$

## The Flash Clock

The flash clock causes any screen attribute cell that has its flash mode enabled to repeatedly swap its foreground and background colours. ZX Spectrum achieves a flash rate of 1.56 Hz, flashing the attribute three times in approximately two seconds, by counting 32 display frames.

The frequency of one display frame is:

$$f_{frame} = \frac{1}{64 \times 10^{-6} \times 312} Hz$$

$$f_{frame} = 50.0801 Hz$$

The frequency of 32 display frames is:

$$f_{frame32} = \frac{50.0801}{32} Hz$$

$$f_{frame32} = 1.565 Hz$$

Put another way, 50.0801 consecutive frames take one second to display, therefore 32 frames take 0.639 seconds, giving the period of one flash cycle.

The counter consists of a five stage ripple counter constructed from D-type flip-flops, as shown in Figure 14-3. Each flip-flop output is configured to toggle at the downward transition of its input clock, with each output successively clocking the next flip flop in line.

The first flip-flop is clocked by V8, which goes low when the electron beam reaches the first row of pixels, once each frame. See Table 11-2. This frame clock is NOR gated by /TCLKB (discussed in Chapter 23, *Hidden Features and Errors*, and which may always be assumed to be low) so the inverted signal /V8 is taken instead.

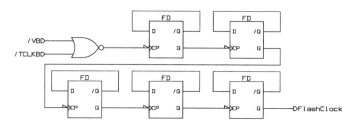

*Figure 14-3: Flash control clock*

# Chapter 15
# Video Addressing

In Chapter 12, *Generating The Display* we saw that the ZX Spectrum display has a resolution of 256 × 192 pixels, with one byte representing eight pixels horizontally; therefore the memory map of the screen is 32 bytes × 192 lines, occupying 6144 bytes of memory.

Along with pixel display bytes the video memory also stores colour information, although at a lower resolution of one colour (attribute) byte for every eight pixel lines in a column. This gives a colour resolution of 32 × 24 colour cells, occupying 768 bytes in memory.

As discussed in Chapter 13, *Video Memory Access*, the video controller reads pixel and attribute bytes in pairs from the video area within the lower 16K DRAM. Due to the limited time with which it has to read the bytes the video controller uses a page-mode read, which allows multiple bytes to be read in quick succession, so long as they are read from the same DRAM row.

As the address bus of DRAM is multiplexed, addresses must be presented to it in two stages, first a memory row address, and second, a memory column address. In the case of the 16K $\mu$PD416 [DS4116] DRAM used in the ZX Spectrum, this means splitting the 14-bit address into two 7-bit addresses, the lower being used for the DRAM row address, the upper for the column. The design implication for the ZX Spectrum page-mode read is that both the display byte and attribute byte addresses must share the same 7-bit row address.

## Address Generation Theory

There are many potential display and attribute byte arrangements that could be considered when designing the video display memory layout, the candidates being limited mainly by timing constraints and the complexity of their implementation. For example, a layout of nine repeating bytes consisting of an attribute byte followed by the eight display bytes it covers, is simple to describe and understand. However, because a display byte will only be, at

most, nine bytes away from its corresponding attribute byte, the least significant four bits of the video address will be different for each byte fetched. This
prevents the lower address bits from being used as a page-mode DRAM row
address for display and attribute bytes, making this proposal unsuitable.

Instead, the 6144 bytes of pixel display data can be stored at the start of the
16K DRAM, followed by the 768 bytes of attribute data. In terms of the CPU
address space, this begins at address 0x4000 (16384 decimal) in the memory map. The video controller on the other hand only has access to this 16K
region of memory, and thus sees it beginning at address zero in its memory
map, thereby simplifying addressing and allowing it to generate an address
between 0x0000 and 0x17FF (6143 decimal) when reading display bytes, and
between 0x1800 and 0x1AFF (6911 decimal) when reading attribute bytes.
The attribute bytes are also at least 768 bytes away from their corresponding
display bytes, allowing a carefully constructed 7-bit DRAM row address to be
shared between them, since $2^7 < 768$.

As described above, the pixels are organised as $192 \times 32$ bytes. These bytes
are read one column after the next, one line after another, in step with the
electron beam as it proceeds left-to-right and down the screen. The position of
the electron beam at any moment is given by the master (horizontal) counter
C8–0 and the vertical line counter V8–0, and it is from these two counters that
the memory address of the video byte to be fetched can be created.

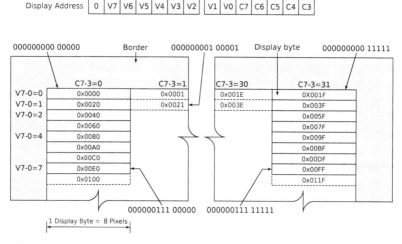

*Figure 15-1: Theoretical display address from horizontal and vertical counters*

The horizontal counter gives the position of the electron beam at a pixel resolution as it scans across the screen, with its origin at the left hand side of the pixel display rectangle. As each fetched pixel display byte returns data for eight pixels at a time, the horizontal counter needs to be divided by eight to give a position in terms of display byte column. Values greater than 31 are not relevant, as they represent periods when the electron beam is within the right, and then left borders. See Table 11-1. Dividing by eight and discarding counts greater than 31 is achieved by considering only five counter bits, $C7{-}3$, since $2^5 = 32$.

The vertical counter gives the position of the electron beam at a pixel row resolution as it scans down the screen, with its origin at the top of the pixel display rectangle. Counter values greater than 191 represent periods when the electron beam is within the bottom, and then top borders. See Table 11-2. A line count of between 0 and 191 can be fully represented by vertical counter bits $V7{-}0$, after discarding $V8$, to give a maximum range of 0 to 255. As this still exceeds 191, further limiting of the count range will be required.

The simplest theoretical addressing scheme is one which starts at zero as the electron beam enters the top left hand corner of the pixel display rectangle, and increments as it moves across each of the 32 eight-pixel display columns for all 192 lines.

Figure 15-1 illustrates such a scheme where the counters are combined into a 14-bit address, the lower significant five bits being supplied by $C7{-}3$, and the most significant bits by $V7{-}0$. Some of the addresses are also shown in their binary form, revealing the contributions made by the two counters.

The address is only valid while the electron beam is within the pixel display rectangle, which is described by the following relationships:

$$C8 = 0 \equiv HBorder = 0$$
$$0 \leq V8..0 \leq 191 \equiv VBorder = 0$$

Thus the address created from these counters will be valid whenever the Border signal is zero; defined in the section called *Vertical Timing* in Chapter 11, *Video Synchronisation*.

We now consider the generation of the attribute byte address. A single attribute byte provides the colour information for eight consecutive display lines within a column, covering an $8 \times 8$ pixel square; thereby reducing the vertical resolution of the display area for an attribute row by a factor of eight. To achieve this reduction, vertical counter bits $V2{-}0$ are excluded from the address generation. As with the display byte address, the least significant five bits are given by the horizontal counter, C7–3, stepping through each of the 32 columns.

Figure 15-2 shows the relationship of the horizontal and vertical counters to the attribute byte address. It should be noted that because the attribute bytes follow the 6144 display bytes in memory, they start at a base address of 0x1800 (6144 decimal), which is easily implemented by pre-loading the most significant bits of the address, A13–10, with binary 0110.

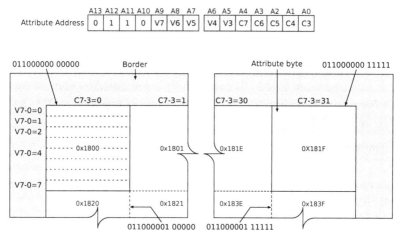

Figure 15-2: Theoretical attribute address from horizontal and vertical counters

Comparing the addresses generated for the display bytes in Figure 15-1 and attribute bytes in Figure 15-2 shows that the lower seven bits of each address, A6–0, differ due to the exclusion of V2–0 from the attribute byte address. See Figure 15-3.

|  | DRAM Column Address | | | | | | | DRAM Row Address | | | | | | |
|---|---|---|---|---|---|---|---|---|---|---|---|---|---|---|
|  | A13 | A12 | A11 | A10 | A9 | A8 | A7 | A6 | A5 | A4 | A3 | A2 | A1 | A0 |
| Attribute Address | 0 | 1 | 1 | 0 | V7 | V6 | V5 | V4 | V3 | C7 | C6 | C5 | C4 | C3 |
| Display Address | 0 | V7 | V6 | V5 | V4 | V3 | V2 | V1 | V0 | C7 | C6 | C5 | C4 | C3 |

Figure 15-3: Comparison of initial theoretical display and attribute addresses

## The ZX Spectrum Addressing

To establish a 7-bit DRAM row address that may be used by both display and attribute byte fetches, the seven least significant address bits common to both addresses are used. These turn out to be the lower seven bits of the attribute address shown in Figure 15-2. The remaining bits of each address create the upper 7-bit DRAM column addresses, where the most significant bits of the attribute byte address are set to 0110 binary, and the display byte address contains V2–0. Figure Figure 15-4 shows both the DRAM row and column addresses for the attribute byte, and three possible options for the display byte DRAM column address.

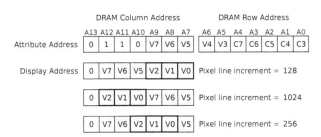

*Figure 15-4: Variations on 7-bit display byte column address*

Because each of the options for the display byte address shown in Figure 15-4 no longer contain vertical counter bits in their order of significance, the display bytes will not be fetched from memory in a linear sequence. The pattern of the fetch sequence depends on where V1–0, which have been moved out of the DRAM row address, are placed within the column address. Eight consecutive lines make up one attribute row, therefore V2–0 should be kept together so that the processor sees a consistent offset between each of these eight lines, as a convenience to software. If V2–0 are placed at A9–7, then the offset becomes $2^7 = 128$ bytes. If placed at A12–10, the offset becomes $2^{10} = 1024$ bytes, and if placed at A10–8 the offset becomes $2^8 = 256$ bytes. A 256 byte offset stands out from the others, as it allows software to step an address from one line to the next simply by incrementing the address high byte, easily achieved with a single Z80 instruction. It is for this reason that the ZX Spectrum ULA implements the third address option.

Analysing this addressing option further, it should be noted that when the processor steps from the eighth to the ninth display line by incrementing the high byte (A16–A8), A11 will become set and A10–8 reset. However, the ninth line is actually addressed when A5 goes to 1, since it is associated with V3; therefore this high byte line stepping trick can only be performed when stepping within an eight pixel high row, defined when A10–8 is not 111 binary.

Generally the processor address of any display byte is calculated by setting A4–0 to the column number, A12–5 to the line number and swapping A10–8 with A7–5. The next line may be stepped to by incrementing the high address byte until A10–8 are 111 binary, at which point the next line address may be found be recalculating the whole address or by applying an address adjustment.

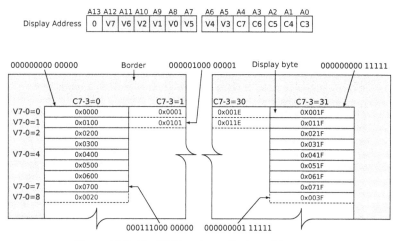

Figure 15-5: The ZX Spectrum display address map

## Generating The Address

The address lines of the ULA are multiplexed to save package pins, and also allows the ULA to connect directly to the 16K DRAM without the need for an external multiplexer. The address that the ULA outputs when performing a video fetch is governed by the following factors:

1. Whether the video controller is about to begin a page mode read.
2. Whether it is about to fetch a display byte.
3. Whether it is about to fetch an attribute byte.

A video fetch occurs when the Border signal is low and C3 is high, as shown by Figure 13-4 and Figure 15-6. Because the DRAM row address must be present on the address bus before /RAS goes low, the ULA must generate the row address early, before C3 goes high. It does this by AND gating C0, C1 and C2 to identify the high transition of C0 prior to C3 going high, adding 142ns to the front of C3's high duration. This produces an address enable signal /AE, which enables the ULA address bus tri-state outputs, passing the address to the DRAM.

$$\overline{AE} = Border + (C3 + (\overline{C0} + \overline{C1} + \overline{C2}))$$

Considering /AE on its own is insufficient to decide whether to produce a row or column address. However, once /RAS has gone low, the row address on the bus must be changed to a column address before /CAS goes low; therefore, while /RAS is high, a row address should be generated (/RSel), and when low, a column address. Additionally, because the video controller is performing a page-mode read of two bytes, two different column addresses need to be generated per row access, one for the display byte (/CDataSel), another for the attribute byte (/CAttrSel).

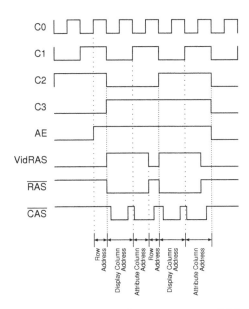

*Figure 15-6: Clock lines C3–0 in relation to RAS and CAS*

Examining Figure 15-6 reveals that C1 is high for the duration of the first

/CAS pulse, responsible for fetching the display byte, and low for the second /CAS pulse, which fetches the attribute byte. Thus by considering C1 along with the /RAS signal, the ULA is able to correctly produce either a row address or one of the two column addresses.

$$\overline{RSel} = VidRAS$$
$$\overline{CDataSel} = \overline{VidRAS} + C1$$
$$\overline{CAttrSel} = \overline{VidRAS} + \overline{C1}$$

These equations show that when VidRAS goes low prior to the second pair of bytes being fetched, a row address will be generated as appropriate. Internally the video generator will always produce one of the three addresses, but will only output them to the address bus while /AE is high.

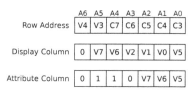

| | A6 | A5 | A4 | A3 | A2 | A1 | A0 |
|---|---|---|---|---|---|---|---|
| Row Address | V4 | V3 | C7 | C6 | C5 | C4 | C3 |
| Display Column | 0 | V7 | V6 | V2 | V1 | V0 | V5 |
| Attribute Column | 0 | 1 | 1 | 0 | V7 | V6 | V5 |

*Figure 15-7: Comparison of ZX Spectrum row and column DRAM addresses*

The ULA generates the required address by combining the row and column address selects, defined above, with the three counter bits given for each of the columns A6 to A0 in Figure 15-7. Where a binary 0 is required, the invert of the appropriate column select is taken on its own, as this being high will force the combining NOR gate for the address line to output 0. Where a binary 1 is required, the select signal is omitted altogether as all inputs of the output NOR will be zero, forcing its output to 1.

$$A0 = \overline{(V5 + \overline{CAttrSel}) + (V5 + \overline{CDataSel}) + (C3 + \overline{RSel})}$$
$$A1 = \overline{(V6 + \overline{CAttrSel}) + (V0 + \overline{CDataSel}) + (C4 + \overline{RSel})}$$
$$A2 = \overline{(V7 + \overline{CAttrSel}) + (V1 + \overline{CDataSel}) + (C5 + \overline{RSel})}$$
$$A3 = \overline{CAttrSel + (V2 + \overline{CDataSel}) + (C6 + \overline{RSel})}$$

$$A4 = \overline{\overline{(V6 + \overline{CDataSel})} + \overline{(C7 + \overline{RSel})}}$$

$$A5 = \overline{\overline{(V7 + \overline{CDataSel})} + \overline{(V3 + \overline{RSel})}}$$

$$A6 = \overline{CAttrSel + CDataSel + \overline{(V4 + \overline{RSel})}}$$

Interestingly, Altwasser did not use the logical choice of C3 for the generation of row address bit A0, as shown in the above equation, but instead used a delayed /C2. To understand why this was necessary, an appreciation of the signal timing involved is required.

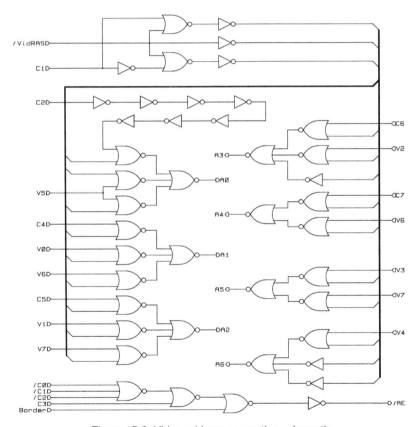

*Figure 15-8: Video address generation schematic*

Referring to Figure 14-1, the video controller delays the output of the serial pixel stream to give itself time to fetch the first pair of display and attribute bytes from memory. As a consequence of this delayed timing, the byte fetches occur one C2 period later than they would have otherwise, and the first /RAS goes low while C3 is high.

Because the generation of A0 requires C3 to be low at this point, /C3 could have been used instead, had it not been for the second byte-pair being fetched immediately after the first, within the same half-period of C3. Thus, the use of C3 or /C3 will always see A0 being set to the same state for every DRAM row address. C2 on the other hand, gives the expected value for A0 at each /RAS pulse.

Nevertheless, it is likely that Altwasser would have suffered additional timing issues with the use of C2, due to the internal propagation delay of the master counter between C3 and C4 (see the section called *The Master Counter* in Chapter 10, *The Internal Clocks*). /RAS is generated from C0 and C1 and will be almost synchronous with C2 and C3 as they change state. C8–4, in comparison, change state after C3–0 and /RAS, and it is this slight delay that keeps C7–4 and V7–0 stable on the address bus while the DRAM responds to /RAS. In contrast, C2 and C3 will be in a state of change as /RAS goes low; therefore their participation in the generation of A0 would lead to an unstable address.

Consequently, C2 is delayed and inverted by seven inverters before being used for the generation of A0, as shown in Figure 15-8. This ensures that C2 changes state after /RAS has been asserted, and more than makes up for the propagation delay within the master counter. The inversion is necessary to correct the state of C2, as the delay causes its previous state to be placed on the address bus. The other counter bits involved in the address generation, C4 and above, do not need to be inverted since they change state on the downward transition of C3, which occurs after the four bytes have been fetched, and therefore during a safe zone.

Ironically, if the same delay and inversion had been applied to C3, then it could have been used in the generation of A0. Either it wasn't obvious that C3 could become a valid candidate, or its high fan-out and consequently slower transition time precluded it from being used reliably.

# Chapter 16
# Analogue Video

Internally the ULA processes colour information as separate red, green, blue and bright signals. As the ZX Spectrum is designed to operate with a television receiver which requires a single composite video signal mixed with a UHF or VHF carrier frequency, the colour and synchronisation components must be combined.

Composite video generation and RF modulation is a complicated analogue process, for which the analogue capability of the 5C000 and 6C000 ULA is not sufficient. However, low cost specialised microchips and modulators were available to perform the task.

The ZX Spectrum uses the National Semiconductor LM1889 to modulate colour signals into a PAL (or NTSC with a different ULA and minor circuit modifications) chroma sub-carrier. This is combined with the raw video output into a composite video signal that is further modulated onto an RF carrier by the UM1233 RF modulator, producing a TV compatible UHF aerial signal.

## Composite Video

A composite video signal combines the three video components of brightness, colour and synchronisation into one, making them simple to transmit. The combination is done in such a way that the receiving television can easily separate the components back out again.

Before the arrival of colour television, a black and white composite video signal consisted of just brightness and synchronisation. In a colour video signal these two components are still present, and are together referred to as luminance or symbolically as Y.

The colour information is added to the luminance component of a composite video signal in a way that makes it invisible to a black and white television. At the introduction of colour, a majority of television programmes were still produced in black and white, and as most viewers owned the cheaper black and white television set, the addition of the colour component had to allow programmes to be viewed by all.

In 1939, Georges Valensi [VALENSI] invented the method of analogue colour encoding used today that allows the transparent transmission of colour information alongside the black and white luminance signal. The mechanism uses colour difference signals, which reduces the amount of information or bandwidth needed to transmit a full colour picture. Together with the luminance Y, there are two colour difference signals, U and V, together referred to as chrominance.

## YUV

The colour difference encoding used in PAL, NTSC and SECAM composite video standards are all based on the YUV colour space, which consists of the black and white luminance Y, and chrominance colour differences U and V.

YUV takes into account the human eyes sensitivity to different colours and this enables it to encode a fixed percentage of each primary colour, significantly reducing the amount colour information that will be transmitted.

The luminance Y is calculated from the weighted sum of RGB colours, with weights reflecting the proportion of each primary colour that gives the luminance of reference white [VIDEODM], according to the ITU-R BT.601 specification:

$$Y = 0.299R + 0.587G + 0.114B$$

The colour difference signals U and V are defined as:

$$U = 0.493(B - Y)$$
$$V = 0.877(R - Y)$$

which, when expressed in terms of RGB, become:

$$U = -0.147R - 0.289G + 0.437B$$
$$V = +0.615R - 0.515G - 0.100B$$

YUV signal generation is described as being lossy because colour information is discarded, and at the television receiver the reconstituted RGB values will never be the same as those before YUV conversion. In the early days of colour television, viewers would not have been aware of this limitation due to the inaccurate picture reproduction of the domestic television. However, the quality of modern television and monitor equipment highlights the inadequacies of such a lossy analogue encoding, and direct analogue RGB or digital signals are now preferred instead.

## PAL Chrominance Modulation

PAL colour encoding amplitude modulates two orthogonal sub-carriers (being 90 degrees out of phase with one another), with the two weighted colour difference signals. This *Suppressed Carrier Quadrature Modulation* produces a single amplitude and phase modulated signal that, when added to the luminance, gives a composite video signal. Being amplitude modulated, the magnitude of the carrier is proportional to the magnitude of the modulating signals, which are therefore said to suppress the carrier. This ensures compatibility with black and white television receivers by confining moments of high carrier amplitude to areas of high colour saturation.

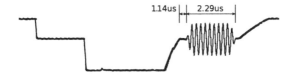

*Figure 16-1: "Colour Burst" synchronisation - 10 cycles of sub-carrier*

At the receiver, the suppressed sub-carrier must be regenerated before the colour difference signals can be demodulated. Usually with quadrature modulation a pilot signal is continuously transmitted to allow the receiver to lock on to the sub-carrier frequency, but with television, such a signal would create visible interference due to the limited bandwidth available. However, unlike most quadrature modulated signals, a television transmission is divided into discrete chunks of information, where a frame consists of 287 repetitions of scan line and horizontal synchronisation information, followed by 25 vertical synchronisation pulses. This makes it possible to provide the receiver with the necessary sub-carrier synchronisation without a continuous pilot, by adding a sine wave "Colour Burst" signal to the video signal during the line blanking period, as shown in Figure 16-1.

Distortions or phase errors in the modulated chrominance signal reduce the orthogonality of the signals and results in crosstalk between the V and U channels. The PAL or "Phase Alternating Line" system designed by Bruch [BRUCH] reverses the phase of the V sub-carrier on alternate lines, so that when the line phase is restored at the receiver, any induced phase error will also be inverted, canceling out errors between adjacent lines. Analogue television receivers use a chrominance delay line to store the colour information for a line so that it may be averaged against the next to remove the phase error. Even though this reduces the vertical colour resolution of the image, it deliberately causes phase errors to manifest themselves instead as subtle changes in saturation, both of which are undetected by the human eye, being far more sensitive to hue and intensity than saturation.

An understanding of how quadrature modulation combines the chrominance signals U and V with a carrier is not required to understand how Y, U and V are generated from RGB. That said, it is important to understand the effect of quadrature modulation on the phase and amplitude of the carrier signal, when considering how to generate the colour synchronisation burst during the horizontal blank. The modulator adds the burst to the video signal, but its location, amplitude, and crucially, its phase are dictated by the YUV generator.

**Quadrature Amplitude Modulation**

Quadrature modulation amplitude modulates two carrier signals with two information signals. One carrier is 90 degrees out of phase with the other, and both are typically sinusoidal. In other words, one carrier is a function of sine, the other a function of cosine. Once modulated, the two signals are summed to produce a single signal that is both phase and amplitude modulated. The modulation and summation is shown diagrammatically in Figure 16-2.

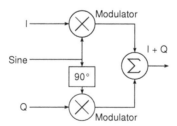

*Figure 16-2: Quadrature amplitude modulator*

Assume that the U signal is modulated by the cosine carrier, and V by the sine carrier. The effect of the modulation on the output carrier can be shown by plotting the vector sum of these two quadrature components, U and V, for an example colour. As V will be modulated 90 degrees out of phase with U, the V and U axes are at 90 degrees to one another, as presented in Figure 16-3.

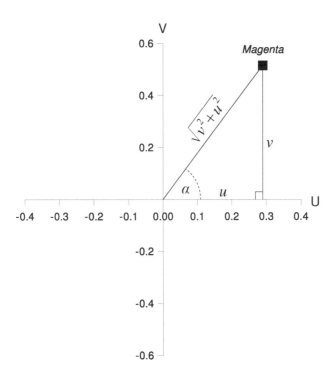

*Figure 16-3: Vector sum of V and U showing phase shift and amplitude*

This diagram shows the vector sum of U and V for magenta, as calculated by the BT.601 equations with values for red and blue of 1. The point plotted forms a right-angle triangle with the U-axis, with its hypotenuse intersecting the origin, such that it describes an angle $\alpha$ with the positive half of the U-axis. This angle gives the phase shift applied to the carrier when modulating U and V for this colour, and the hypotenuse gives the carriers amplitude.

All angles within the vector diagram are measured with respect to the positive U-axis, and proceed in counter-clockwise direction.

The amplitude of modulated carrier for magenta in Figure 16-3 is given by:

$$amplitude = \sqrt{v^2 + u^2} = \sqrt{0.515^2 + 0.290^2} = 0.591$$

The phase shift of modulated carrier in degrees is given by:

$$\alpha = sin^{-1}\left(\frac{v}{\sqrt{v^2 + u^2}}\right)$$

$$\alpha = sin^{-1}\left(\frac{0.515}{0.591}\right) = 60.6$$

Thus, magenta has a modulated carrier phase shift of 60.6 degrees and a carrier amplitude scale factor of 0.591.

## ULA Analogue YUV Generation

Generation of luminance and chrominance signals requires a digital to analogue converter (DAC) that can sum appropriately weighted values for each of the red, green and blue digital inputs.

The luminance signal conveys intensity and synchronisation, and will have a value that ranges from zero for sync, a larger positive value for black through to some higher value for bright white. The chrominance signals U and V, by comparison, have values that are zero for both black *and* white, as these colours are represented by their luminance component alone. This can be demonstrated by plotting the vector sum of the quadrature components, U and V, for each of the primary colours and their complements, as presented in Figure 16-5.

The simplest digital to analogue converter is constructed from a network of single and double valued resistors, arranged into an array called an R-2R resistor ladder. See in Figure 16-4. Each resistor tap along the length of the array provides an input bit supplying 0 or Vref volts, depending on the state of the bit.

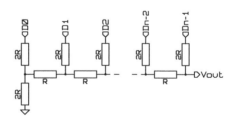

*Figure 16-4: R-2R resistor ladder*

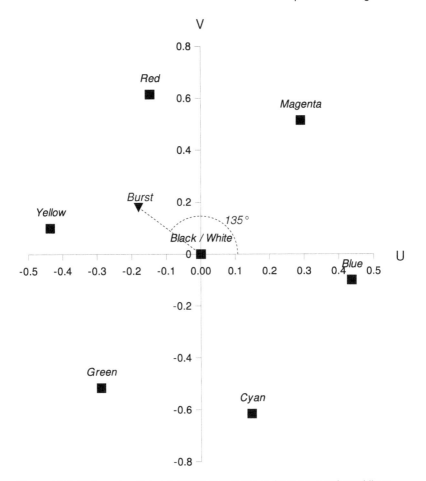

*Figure 16-5: PAL vector diagram of UV colour space for even numbered lines*

If the input bits are numbered from 0, an arbitrary bit *i* contributes the following to the output voltage:

$$V out = \frac{Vref}{2^{i+1}}$$

The full scale output voltage for a N-bit ladder is given by:

$$V out = V ref \sum_{i=0}^{N-1} \frac{1}{2^{i+1}}$$

The R-2R ladder is not affected by the value of R, and as any resistance drift during operation would affect all resistors, the output voltage remains stable. Even so, despite the simplicity and stability of the resistor ladder, it is not suitable for the generation of non-linear weighted voltages and difficult to implement within a Ferranti ULA. The CML ULA has limited analogue capabilities provided by its peripheral cells, as they are intended for digital interfacing, and do not contain the number of resistors necessary to construct an R-2R resistor ladder. They do, however, contain a useful range of resistors in various ratios.

The colour weight coefficients of the three YUV equations may be expressed as ratios of one another. If these ratios are duplicated across a set of resistors making up one half of a voltage divider, then by switching combinations of these resistors in and out of the divider, the output voltage can be made to vary in proportion to the resistor ratios, and therefore the colour weights.

This is the technique used in the ZX Spectrum ULA, creating a simple DAC implemented with a limited number of transistors and resistors having limited interconnection possibilities. The design combines resistors in both parallel and serial configurations to achieve the values required, and incorporates a number of features that provide signal gain and temperature stability, preventing the brightness from shifting as the IC heats up.

**General Circuit Overview**

The YUV signal generation consists of three analogue circuits which are based on the same core design, driven by a logic circuit that provides the necessary RGB signals and timing.

Altwasser prototyped the YUV circuits using discrete transistors and found that when blue alone was displayed, it was very dark and hardly visible on a number of televisions. To address this he increased the blue coefficient of the luminance equation:

$$Y = 0.299R + 0.587G + 0.151B$$

Each of the YUV circuits consist of an output transistor whose output voltage depends on the combination of resistors connected to its emitter. These resistors are switched in and out of the circuit depending on the combinations of red, green and blue signal present at the input, and their values chosen to match the ratio of weight coefficients in the corresponding luminance and colour difference equations. A similarly weighted burst signal complements the colour signals in the chrominance U and V circuits, controlling the synchronisation colour burst pulse they generate.

The following analysis assumes a negligible transistor base current and that the switching transistors either cut-off or saturate, depending on their control input.

### Luminance Y Generation

The luminance circuit takes single bit, red, green and blue inputs and produces an output voltage from the weighted sum of colour inputs. The luminance coefficients determine how significant the contribution of each input colour is to the overall brightness of the output; therefore blue is given the least significant bit, and green the most.

When viewed on a monochrome display, each of the eight possible colours increases in lightness across a grey scale from black to white. When a colour is to be produced with highlight (or brightness) turned on, the output voltage is increased by a small amount. To reduce the complexity of the design, the ULA generates an inverted luminance signal, /Y, which is reverted by the composite video circuit on the ZX Spectrum PCB.

The luminance circuit for the 6C001E ULA is shown in Figure 16-6. It consists of an output transistor, Q3, connected to the ULA pin via an emitter-follower transistor, Q4, that buffers Q3 from a large external circuit load and provides a low impedance output.

The voltage output by Q4 through /Y, as measured across the external load resistor RL, tracks the voltage at its base, with a base-emitter voltage drop $V_{BE}$ of 0.7v. The voltage at the base of Q4 is determined by the voltage drop across R2, which is controlled by the remainder of the circuit.

Q2 is the control transistor that establishes the operating point of Q3. Q1 is always on, switching R3 into the emitter of Q2, and for normal colour brightness levels, Q5 is also on, adding R4 in parallel with R3. The total emitter resistance is therefore:

$$\frac{1}{RE_{Q2}} = \frac{1}{3100} + \frac{1}{7700}$$

$$RE_{Q2} = 2210R$$

This 2210 ohm resistance in series with R1 gives 13010 ohms across 5v, minus the Q2 base-emitter drop of 0.7v. From Ohm's Law, the emitter current $IE_{Q2}$ is given by:

$$IE_{Q2} = \frac{5 - 0.7}{2210 + 10800} = 0.331 \times 10^{-3} A$$

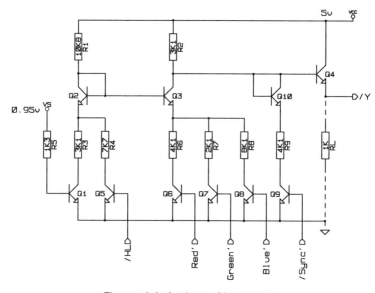

Figure 16-6: Analogue /Y generation

The current produces a voltage drop across the emitter resistance, setting the voltage at the emitter:

$$VE_{Q2} = 0.331 \times 10^{-3} \times 2210 = 0.730v$$

As Q2 and Q3 both have their bases tied to the same low impedance point, the voltage at the emitter of Q2 (0.730v) is reflected at the emitter of Q3.

The voltage across R2, which dictates the overall output voltage, is proportional to the current flowing through it, and is determined by which of the Red', Green' and Blue' transistors, Q6-Q8 are on. The resistor values they switch into the emitter circuit are chosen to produce currents in the same ratio as the luminance coefficients.

As current is inversely proportional to resistance, the resistance ratios are like-wise inverted:

$$0.299 : 0.587 \approx 2100 : 4100$$
$$0.589 : 0.151 \approx 8100 : 2100$$

In the 6C001 ULA, 8K1 resistor is formed from two 4K1 resistors in series.

As the voltage at the emitter of Q3 is known and fixed, it is possible to calculate the currents drawn by each of the emitter resistors:

$$I_{Red} = \frac{0.730}{4100} = 0.178 \times 10^{-3} A$$

$$I_{Green} = \frac{0.730}{2100} = 0.348 \times 10^{-3} A$$

$$I_{Blue} = \frac{0.730}{8100} = 0.092 \times 10^{-3} A$$

Different combinations of red, green and blue input results in different currents being drawn through R2 by the emitter resistors. During the active region of the display, /Sync will be high and Q9 on, causing R9 and Q10 to draw addition current through R2. R9 in series with R2 gives 7200 ohms across 5v, minus the base-emitter drop of 0.7v, giving a current draw of:

$$I_{SYN} = \frac{5 - 0.7}{7200} = 0.597 \times 10^{-3} A$$

This current must also be taken into account when calculating the voltage drop across R2 for each colour. White for example, which is created when Q6-Q9 are on, draws a total of $1.213 \times 10^{-3}$A through R2, causing it to experience a voltage drop of:

$$R2(white)_V = 1.213 \times 10^{-3} \times 3100 = 3.76v$$

The output produced by Q4 for white is the voltage at its base, being 5v minus the drop across R2, less the base-emitter voltage drop:

$$\overline{Y_{white}} = 5 - 3.76 - 0.7v = 0.54v$$

**Highlight Mode**

When the highlight mode is enabled, /HL is low, switching off Q5 and increasing $RE_{Q2}$ to 3100 ohms (the value of R3), causing an associated decrease of $IE_{Q2}$:

$$IE_{Q2(bright)} = \frac{5 - 0.7}{3100 + 10800} = 0.309 \times 10^{-3} A$$

This leads to an increase in the voltage at the emitters of Q2 and Q3:

$$VE_{Q2(bright)} = 0.309 \times 10^{-3} \times 3100 = 0.959v$$

Consequently, the currents drawn by each of the red, green and blue resistors are proportionally higher. Note that no more than $13.035 \times 10^{-4}$A may be drawn through R2, since the voltage at the collector of Q3 cannot go below 0.959v. At this threshold there is no longer any voltage drop across Q3, which will be in saturation; therefore the generation of bright white drops the maximum 4.041v across R2. The circuit currents, voltage drop across $V_{R2}$ and /Y output are given in Table 16-1 for all 16 possible colours.

| Colour | IGreen | IRed | IBlue | /Isync | R2v | /Y |
|---|---|---|---|---|---|---|
| Black | | | | 0.597 | 1.851 | 2.449 |
| Blue | | | 0.090 | 0.597 | 2.131 | 2.169 |
| Red | | 0.178 | | 0.597 | 2.404 | 1.896 |
| Magenta | | 0.178 | 0.090 | 0.597 | 2.683 | 1.617 |
| Green | 0.348 | | | 0.597 | 2.930 | 1.370 |
| Cyan | 0.348 | | 0.090 | 0.597 | 3.209 | 1.091 |
| Yellow | 0.348 | 0.178 | | 0.597 | 3.482 | 0.818 |
| White | 0.348 | 0.178 | 0.090 | 0.597 | 3.761 | 0.539 |
| Bright Black | | | | 0.597 | 1.851 | 2.449 |
| Bright Blue | | | 0.118 | 0.597 | 2.218 | 2.082 |
| Bright Red | | 0.234 | | 0.597 | 2.576 | 1.724 |
| Bright Magenta | | 0.234 | 0.118 | 0.597 | 2.944 | 1.356 |
| Bright Green | 0.457 | | | 0.597 | 3.267 | 1.033 |
| Bright Cyan | 0.457 | | 0.118 | 0.597 | 3.634 | 0.666 |
| Bright Yellow | 0.457 | 0.234 | | 0.597 | 3.992 | 0.308 |

| Colour | IGreen | IRed | IBlue | /ISync | R2v | /Y |
|---|---|---|---|---|---|---|
| Bright White | 0.457 | 0.234 | 0.118 | 0.597 | 4.041 | 0.259 |
| Sync | | | | | 0.00 | 4.3 |

*Table 16-1: Luminance circuit currents (mA), voltages and output levels*

## Synchronisation

The horizontal and vertical synchronisation pulses are added to the luminance signal by removing the voltage offset applied to the luminance output when sync is not being generated. This causes the output to go above the reference black level of the inverted luminance signal, as required for synchronisation pulses in an inverted luminance signal.

Horizontal and vertical synchronisation occurs during a period of line blanking, when active region of the display is not being generated; therefore the red, green and blue signals will be off, and the current drawn through R2 a result of Q9 being on, providing the black offset. During the sync pulse, /Sync goes low, shutting down Q9 and the associated current in R2. The base of Q4 is thus pulled up to the 5v rail, and 4.3v measured across the load resistor in the emitter of Q4. During vertical synchronisation, red, green and blue will again be off, and /Sync will go low for four complete scan lines.

## Temperature Stability

The luminance circuit contains features that compensate for transistor fabrication variances and temperature effects. As the ULA gets hot, the base-emitter voltage drop will decrease for all transistors. The voltage at the emitter of Q2 will rise as a result, increasing the current drawn through R2 and the voltage dropped across it, which reduces the voltage at the base of Q4. The lower base-emitter voltage drop of Q4 increases the voltage at its emitter by more than the voltage reduction experienced at its base, and so Q10 is included to offset this.

The reduced base-emitter voltage of Q10 increases the current drawn through R2 and further lowers the voltage at the base of Q4. This does not achieve complete cancellation of temperature effects seen at the emitter of Q4, but approximately halves the voltage gain.

## Chrominance U Generation

The U chrominance circuit has the same configuration as the luminance circuit, with Q2 controlling the emitter voltage of the output transistor, Q3. Line
and frame synchronisation are not carried by the colour difference signals,
but they do indicate where the chrominance modulator will insert the colour
sub-carrier synchronisation burst, and what phase it will be.

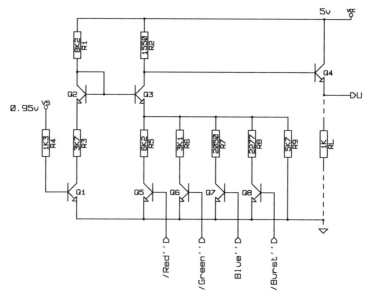

*Figure 16-7: Chrominance U generation*

Technically an inverted signal is produced by the ULA, but once modulated
with the similarly inverted V chrominance signal, a modulated sub-carrier is
produced with the correct magnitude and phase when measured with respect
to the colour burst; therefore the notion that the signal is inverted is dropped
so that it becomes referred to as U.

Figure 16-7 shows the chrominance U generation circuit. The collector and
emitter resistors of Q2 see a current of $0.361 \times 10^{-3}$A being developed through
them:

$$IE_{Q2} = \frac{5 - 0.7}{8200 + 3700} = 0.361 \times 10^{-3}A$$

This current produces a voltage drop across R3 that is reflected at the emitters of Q2 and Q3:

$$VE_{Q2} = 0.361 \times 10^{-3} \times 3700 = 1.337v$$

The current drawn through R2, which dictates the overall output voltage, is determined by which of the /Red", /Green", Blue" and /Burst" transistors (Q5-Q8) are on. The resistor values they switch into the emitter circuit are chosen to match the ratio of chrominance U coefficients, but are inverted since current is inversely proportional to resistance:

$$0.147 : 0.289 \approx 3100 : 6200$$
$$0.289 : 0.437 \approx 2050 : 3100$$

By definition the chrominance U signal can go positive or negative with respect to zero, as shown by the vector sum diagram of Figure 16-5. NPN transistors will only switch positive voltages between 0 and the supply rail, here 5v, so for the signal to represent a negative value, a zero signal level DC offset must be calculated by considering how negative the output signal may become. The LM1889 colour encoder on the ZX Spectrum PCB compares the U and V chrominance signals against this reference DC offset to determine whether or not they represent a negative value. See the section called *LM1889 Modulation Circuit*.

The chrominance U equation states that the red and green components make a negative contribution to the overall signal level; therefore by adding them to the output when they are inactive effectively results in them being subtracted when they are active. The circuit design simply takes inverted signals for red and green so that their switching transistors Q5 and Q6 are off whenever their colour is present.

The currents drawn through R2 for each of the three colour signals are therefore:

$$I_{\overline{Red}} = \frac{1.337}{6200} = 0.216 \times 10^{-3} A$$

$$I_{\overline{Green}} = \frac{1.337}{3100} = 0.431 \times 10^{-3} A$$

$$I_{Blue} = \frac{1.337}{2050} = 0.652 \times 10^{-3} A$$

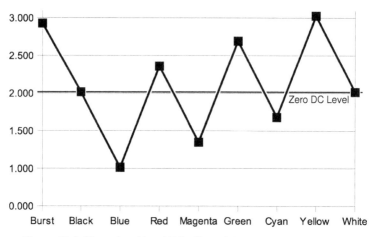

*Figure 16-8: Zero signal level DC offset of the chrominance U signal*

The colour burst control signal, which is discussed fully in the section called *Burst Generation*, also makes a negative contribution to the U output; therefore its weight is included in the output total whenever /Burst[II] is inactive, drawing $0.587 \times 10^{-3}$A through R2:

$$I_{\overline{Burst}} = \frac{1.337}{2277} = 0.587 \times 10^{-3} A$$

Finally, some adjustment of U is required to align its value of white and black with that of chrominance V. This is performed by introducing a fixed current draw through R2 via R9:

$$I_{Fixed} = \frac{1.3375}{5700} = 0.235 \times 10^{-3} A$$

For example, when magenta is sent to the television during the active part of the display, Q6 to Q8 will be on, and Q5 off, drawing $1.905 \times 10^{-3}$A through R2:

$$I_{magenta} = 0.431 \times 10^{-3} + 0.652 \times 10^{-3} + 0.587 \times 10^{-3} + 0.235 \times 10^{-3}$$
$$= 1.905 \times 10^{-3} A$$

This produces a voltage of 2.95v across R2:

$$R2(magenta)_V = 1.905 \times 10^{-3} \times 1550 = 2.95v$$

The output produced by Q4 is the voltage at its base, which is 5v minus the drop across R2, minus the base-emitter voltage drop:

$$\overline{U_{magenta}} = 5 - 2.95 - 0.7v = 1.35v$$

The chrominance U output for each of the eight colours and the colour burst are similarly derived by summing the relevant R2 currents and calculating the voltage drop across R2. These are presented in Table 16-2. Note that the RGB inputs for black, including the burst period, are mapped to those of white to ensure that the quadrature modulation of U and V for both white and black produce the same zero phase and amplitude modulation of the colour sub-carrier. This mapping is explained in the section called *YUV Control Signals*.

| Colour | /I$_{Green}$ | /I$_{Red}$ | I$_{Blue}$ | /I$_{Burst}$ | R2$_v$ | U |
|---|---|---|---|---|---|---|
| Black | | | 0.090 | 0.587 | 2.284 | 2.016 |
| Blue | 0.348 | 0.178 | 0.090 | 0.587 | 3.287 | 1.013 |
| Red | 0.348 | | | 0.587 | 1.942 | 2.358 |
| Magenta | 0.348 | | 0.090 | 0.587 | 2.953 | 1.347 |
| Green | | 0.178 | | 0.587 | 1.608 | 2.692 |
| Cyan | | 0.178 | 0.090 | 0.587 | 2.619 | 1.681 |
| Yellow | | | | 0.587 | 1.273 | 3.027 |
| White | | | 0.090 | 0.587 | 2.284 | 2.016 |
| Burst | | | 0.090 | | 1.374 | 2.926 |

*Table 16-2: Chrominance U circuit currents (mA), voltages and output levels*

**Chrominance V Generation**

The chrominance V generation is very much like that of chrominance U, but has the added complexity of inverting the phase of the signal on alternate lines. To control this inversion the ULA generates a timing signal that, when low, inverts the RGB signals and reverses the polarity of the Burst signal. See the section called *YUV Control Signals*.

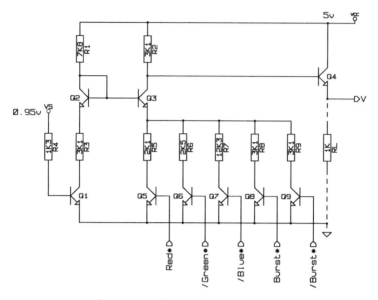

*Figure 16-9: Chrominance V generation*

Figure 16-9 shows the chrominance V generation circuit. The collector and emitter resistors of Q2 see a current of $0.309 \times 10^{-3}$A being developed through them:

$$IE_{Q2} = \frac{5 - 0.7}{10800 + 3100} = 0.309 \times 10^{-3} A$$

This current produces a voltage drop across R3 that is reflected at the emitters of Q2 and Q3:

$$VE_{Q2} = 0.309 \times 10^{-3} \times 3100 = 0.959v$$

The voltage across R2, which dictates the overall output voltage, is determined by which of the Red*, /Green*, /Blue*, Burst* and /Burst* transistors (Q5-Q9) are on. Once again, the resistor values switched into the emitter circuit are chosen to match the inverse ratio of chrominance V coefficients, given that current is inversely proportional to resistance:

$$0.615 : 0.515 \approx 2502 : 2100$$
$$0.515 : 0.100 \approx 12300 : 2100$$

As with the chrominance U signal, V can go positive or negative with respect to zero. Thus for the signal to represent a negative value, a zero signal level DC offset must be established.

The currents drawn through R2 for each of the three colour signals are therefore:

$$I_{Red} = \frac{0.959}{2100} = 0.457 \times 10^{-3}A$$

$$I_{\overline{Green}} = \frac{0.959}{2502} = 0.383 \times 10^{-3}A$$

$$I_{\overline{Blue}} = \frac{0.959}{12300} = 0.078 \times 10^{-3}A$$

In addition to the RGB colour signals, the colour burst control, which is discussed fully in the section called *Burst Generation*, also makes a contribution to the V output during the active region of the display, as /Burst* will be high and Burst* low; therefore $I_{/Burst*}$ is included in the output total whenever the colour burst is not being generated, drawing an additional $0.309 \times 10^{-3}$A through R2:

$$I_{\overline{Burst}} = I_{Burst} = \frac{0.959}{3100} = 0.309 \times 10^{-3}A$$

When the colour burst is being generated, either $2 \times I_{Burst}$ or $0 \times I_{Burst}$ contributes to the current drawn through R2, depending on whether an even or odd numbered scan line is being produced. For further information see the section called *Burst Generation*.

For example, magenta is sent to the television during the active part of the display, Q5, Q6 and Q9 will be on, drawing $1.149 \times 10^{-3}$A through R2. This produces a voltage of 3.56v across R2:

$$R2(magenta)_V = 1.149 \times 10^{-3} \times 3100 = 3.56v$$

The output produced by Q4 is the voltage at its base, which is 5v minus the drop across R2, minus the base-emitter voltage drop:

$$\overline{V_{magenta}} = 5 - 3.56 - 0.7v = 0.74v$$

During odd numbered scan lines, numbering from 1, the single bit RGB input signals are inverted to produce an inverted V output, discussed in the section called *YUV Control Signals*. Both the true and inverted output levels are shown in Table 16-3. Note that the RGB inputs for black, including the burst period, are mapped to those of white to ensure that the quadrature modulation of U and V for both white and black produce the same zero phase and amplitude modulation of the colour sub-carrier. This mapping is explained in the section called *YUV Control Signals*.

| Colour | /I$_{Green}$ | I$_{Red}$ | /I$_{Blue}$ | I$_{Burst}$ | R2$_V$ | V |
|---|---|---|---|---|---|---|
| Black | | 0.457 | | 0.309 | 2.375 | 1.925 |
| Blue | 0.383 | | | 0.309 | 2.147 | 2.153 |
| Red | 0.383 | 0.457 | 0.078 | 0.309 | 3.805 | 0.495 |
| Magenta | 0.383 | 0.457 | | 0.309 | 3.563 | 0.737 |
| Green | | | 0.078 | 0.309 | 1.201 | 3.099 |
| Cyan | | | | 0.309 | 0.959 | 3.341 |
| Yellow | | 0.457 | 0.078 | 0.309 | 2.616 | 1.684 |
| White | | 0.457 | | 0.309 | 2.375 | 1.925 |
| Burst | | 0.457 | | 0.618 | 3.334 | 0.966 |
| Black Inverted | 0.383 | | 0.078 | 0.309 | 2.389 | 1.911 |
| Blue Inverted | | 0.457 | 0.078 | 0.309 | 2.616 | 1.684 |
| Red Inverted | | | | 0.309 | 0.959 | 3.341 |
| Magenta Inverted | | | 0.078 | 0.309 | 1.201 | 3.099 |
| Green Inverted | 0.383 | 0.457 | | 0.309 | 3.563 | 0.737 |
| Cyan Inverted | 0.383 | 0.457 | 0.078 | 0.309 | 3.805 | 0.495 |
| Yellow Inverted | 0.383 | | | 0.309 | 2.147 | 2.153 |
| White Inverted | 0.383 | | 0.078 | 0.309 | 2.389 | 1.911 |
| Burst Inverted | 0.383 | | 0.078 | | 1.188 | 3.112 |

*Table 16-3: Chrominance V circuit currents (mA), voltages and output levels*

The vector sum of U with V for an inverted scan line produces the vector diagram shown in Figure 16-10.

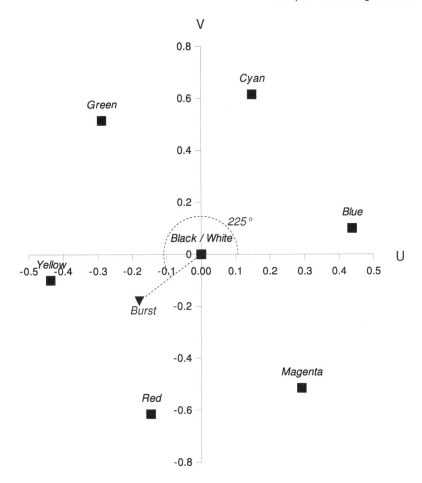

*Figure 16-10: PAL vector diagram of UV colour space for odd numbered lines*

## Burst Generation

For a PAL video signal, the colour burst is generated by setting U and V to values that produce a phase shift of the colour sub-carrier, generated by the LM1889 modulator, of 135 degrees (225 degrees on inverted V lines). This sub-carrier burst lasts for $2.29\mu s$, approximately 10 cycles, and is added to the luminance signal during the blanking period, following the horizontal sync.

A burst phase shift of 135 degrees is produced by a negative U and positive V of the same magnitude. A phase shift of 225 degrees is produced by a negative U and V of the same magnitude. Compare Figure 16-5 and Figure 16-10.

During the colour burst the colour output multiplexer will be producing black, therefore transistors Q6 and Q7 of the chrominance U generator will be on, with Q8 off. For the colour burst enable signal to produce a negative U, it must contribute to the overall voltage offset when it is not active. The colour burst enable is therefore an active low signal, /Burst[II], which removes the burst current component when it is enabled.

The colour burst component of the chrominance V signal is more complicated in its generation. For even numbered lines, chrominance V must include a positive burst component to produce a phase shift of 135 degrees. For odd numbered lines, it must include a negative burst component. The chrominance V generator manages this by using two burst control signals, Burst* and /Burst*.

During the active region of the display, signals /Burst* and Burst* are high and low respectively, drawing a consistent $0.309 \times 10^{-3}$A through R2. When generating the burst period for odd numbered lines, both Burst* and /Burst* are deactivated, removing the burst current completely from that drawn through R2. For even numbered scan lines both Burst* and /Burst* are activated for the burst period, doubling the burst current drawn through R2. This is illustrated in Figure 16-11, and demonstrated in Table 16-3.

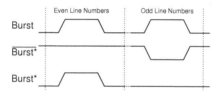

*Figure 16-11: V Burst\* and /Burst\* for odd and even lines*

The burst and chrominance V timing signals are created by the circuit shown in Figure 16-12. Burst is generated from the combination of /C8, /C7, C6–4, producing a signal that is active for 16 pixel clock transitions or $2.29\mu s$. The PAL specification defines the period between the rising edge of the horizontal sync and the colour burst as the breezeway, and gives it a reference duration of $0.9\mu s$. The 5C ULA generates a breezeway of $2.29\mu s$, which is more than double the specification value and is known to cause picture stability problems with Hitachi and Grundig televisions [SM48]. Consequently the breezeway timing was shortened to $1.14\mu s$ with the 6C ULA by delaying the horizontal sync (see the section called *Horizontal Synchronization* in Chapter 11, *Video Synchronisation*).

| Description | Cycle Start | Cycle End | C8–0 at Start |
|---|---|---|---|
| Blanking Period | 320 | 415 | 101 000 000 |
| Horizontal Sync | 336 (5C) | 367 (5C) | 101 010 000 (5C) |
| | 344 (6C) | 375 (6C) | 101 011 000 (6C) |
| Colour Burst | 384 | 399 | 110 000 000 |

*Table 16-4: Burst period within horizontal blanking*

Deciding whether an even or odd scan line is being generated is achieved by considering the lowest bit of the vertical scan line counter, V0, at the start of each new scan line, as defined by the horizontal sync. V0 cannot be used directly as the odd/even timing signal, as it changes state when the electron beam passes the left hand edge of the pixel display rectangle. As scan line numbering is purely arbitrary, Altwasser defined an even scan line as being one where V0 equals zero at the horizontal sync. This effectively numbers scan lines from one, and maintains the convention of inverting the chrominance V for odd scan lines.

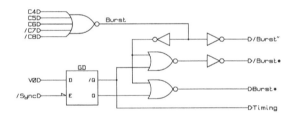

*Figure 16-12: PAL V burst and control signal timing*

The circuit shown in Figure 16-12 takes V0 and stores it in a gated D transparent latch, triggered by the composite /Sync. The complement of the latch output is taken as the odd/even scan line indication signal, referred to as *Timing* by the ZX Spectrum display generation patent [ALTWASSERDC], so that the current scan line is considered even when it is high.

In addition, two chrominance V burst signals are generated, Burst* and /Burst*. When Timing is high, no inversion of the chrominance V will be performed, and therefore /Burst* is held high for the duration of the scan line, with Burst* going high during the colour burst period. When Timing is low, chrominance V will be inverted, and therefore Burst* is held low for the duration of the scan line, with /Burst* going low during the colour burst period. This forces the chrominance V generator to create a positive going burst component for even scan lines, and a negative going burst for odd scan lines. Figure 16-11 illustrates the state of the burst signals for odd and even lines.

| Timing | RGB | Burst Output |
|:------:|:-------:|:------------|
| 1 | Normal | < DC offset |
| 0 | Inverted | > DC offset |

*Table 16-5: Effect of the PAL odd/even Timing signal on RGB and Burst*

**YUV Control Signals**

The YUV signal generation circuits each require different forms of the red, green and blue colour signals, in addition to various timings of burst signal.

The RGB signals from the colour output multiplexer are duplicated three times and processed specifically for each of the Y, U and V generation circuits. In particular, the circuits processing the RGB signals for the U and V generators monitor the input looking for the colour black. When present, the RGB signals are forced high to represent white before the relevant RGB signals are inverted, as required by their intended U and V generator.

Referring to Figure 16-13, the /Y generator takes the RGB, HL and /Sync signals straight from the colour output multiplexer, inverting HL before it is passed to the /HL input of Figure 16-6. These signals are labeled $Red^I$, $Green^I$, $Blue^I$ and /HL. Buffers are added to these signals to align them with the slightly delayed input signals of the U and V generators.

For the U generator, the RGB signals from the colour output multiplexer are NOR gated together to create the $BLACK^{II}$ signal that goes high whenever black is being produced. This black signal is NOR gated back with the RGB signals to force them to produce white when black is present, forming signals $/Red^{II}$, $/Green^{II}$ and $/Blue^{II}$. $/Blue^{II}$ is inverted to give $Blue^{II}$ before being passed to the U generator.

The Red, Green and Blue signals intended for the chrominance V generator are first XNOR gated with the Timing signal to create an RGB signal set that is inverted for even numbered scan lines. These signals are then NOR gated to create the $BLACK^*$ signal that goes high whenever black is being generated. This black signal is NOR gated back with the XNOR gated RGB signals to force them to produce white when black is present, forming signals $/Red^*$, $/Green^*$ and $/Blue^*$. $/Red^*$ is inverted to give $Red^*$ before being passed to the V generator.

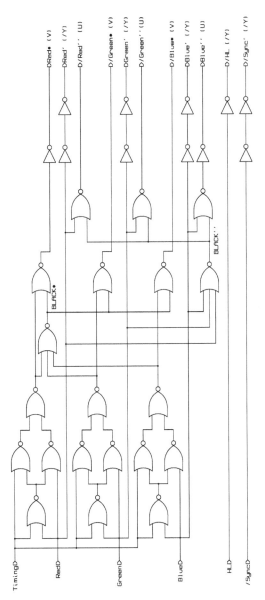

*Figure 16-13: YUV red, green and blue control signals*

## NTSC Chrominance Modulation

NTSC chrominance modulation is very similar to that of PAL, and the ITU-R BT.601 specification of YUV used in the ZX Spectrum is compatible with both video formats.

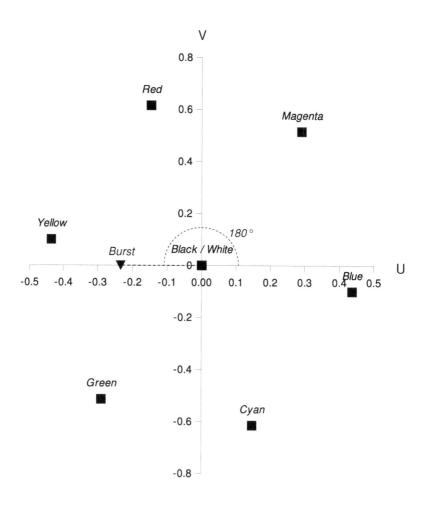

*Figure 16-14: NTSC vector diagram showing UV colour space*

The NTSC video standard defines a colour sub-carrier frequency for colour transmission of approximately 3.58 MHz, and does not invert the phase of the chrominance V component on alternate scan lines. Television pictures in this format also contain fewer scan lines per frame, which, at $64\mu s$ per line, produce 60 frames per second.

The colour burst phase for an NTSC video signal is 180 degrees, which means that there is no burst component of the chrominance V signal. See Figure 16-14.

The 6C011E ULA generates an NTSC compatible colour signal by removing the burst component from the chrominance V signal, and disabling the inversion of V on odd scan lines. It does this by ensuring that the Timing signal is always high, and that both Burst[*] and /Burst[*] are held inactive.

Figure 16-15 illustrates the simplest design change that achieves this. The matrix cells for the gated D transparent latch are interconnected such that the Q and /Q output signal tracks cross over one another, making it a simple modification to join them together and disconnect them from latch. This combined signal would then be pulled high by one of the unused matrix cell load resistors, forcing Burst[*] to be low and both /Burst[*] and Timing to be high.

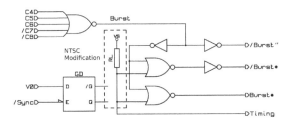

*Figure 16-15: NTSC V burst and control signal timing*

## LM1889 Modulation Circuit

The circuit presented in Figure 16-16 shows a simplified configuration of the LM1889 colour sub-carrier modulator and composite video generator on the ZX Spectrum PCB, in this case for the issue 2 model. It is provided here to illustrate how the YUV signals are processed and combined to form the composite video signal passed to the UM1233 RF modulator. The LM1889 generates a PAL colour sub-carrier from a 4.4336 MHz crystal, and an NTSC colour sub-carrier from a 3.579545 MHz crystal.

A voltage divider comprised of R40 and R41 sets the zero signal level for the chrominance modulator to 4.8v. U or V signals dropping below this value will be considered negative. Variable resistors VR1 and VR2 form part of two voltage dividers that allow the zero signal levels of the ULA U and V chrominance output to be adjusted to 4.8v.

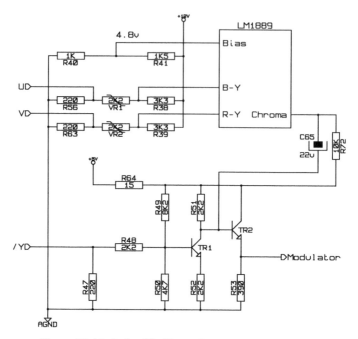

*Figure 16-16: A simplified issue 2 composite video circuit*

TR1 operates in a collector-follower mode, and therefore inverts and amplifies the luminance signal /Y before passing it to the emitter-follower output stage, TR2.

The modulated chroma sub-carrier is AC coupled to the output stage of the modulator driver, superimposing the colour signal onto the reverted luminance.

Later issues of the ZX Spectrum use a more elaborate U and V chrominance bias circuit that does not require setting manually. A simplified issue 2 version is presented here as it is the easiest to understand.

# Chapter 17
# CPU Memory Access

The ULA generates control signals for the ROM and lower 16K dynamic RAM on behalf of the CPU, and combines them with those it generates while performing a video update. External circuitry on the ZX Spectrum circuit board generates the memory control signals for the (optional) upper 32K RAM [1]. See Chapter 8, *The Memory Map*, for further details on the ZX Spectrum's memory arrangement, and the section called *Dynamic RAM* in Chapter 3, *The Standard Microcomputer*, for an overview of dynamic memory devices.

The CPU provides a 16-bit address bus that must be multiplexed before it can be connected to dynamic RAM. Dynamic RAM address bus multiplexing takes a complete set of signals and divides them in two, so that only one half of the bus is available at any one time. A select signal determines which half of the address bus the multiplexer is outputting.

The DRAM and multiplexer control signals are directly influenced by the state of the Z80 memory control signals and address bus, and a full understanding of these is necessary to understand the CPU DRAM interface.

## Z80 CPU Read and Write Cycle

A Z80 instruction consist of a number of M (machine) cycles, each of which consist of between four and six clock cycles, or T-states. The first M-cycle, or M1, is the instruction fetch cycle during which the Z80 loads the next instruction to be executed from memory. Further M-cycles may follow if it is a multi-byte Op Code, or to complete the memory and I/O activity required by the instruction.

The Z80 CPU provides a number of control signals through which it accesses memory devices. The Z80 was designed by Zilog Inc. in 1976, a time when dynamic RAM was much cheaper than static RAM. Zilog engineers deliberately arranged for the Z80s signal timing to closely match those of dynamic RAM, simplifying the interface and giving it the edge on its competitors. It even went as far as to provide an on-chip refresh controller, removing the additional components required by other CPUs. See the section called *Dynamic RAM* in Chapter 3, *The Standard Microcomputer* for further information on dynamic RAM refresh.

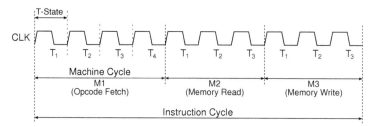

*Figure 17-1: Z80 machine and time state relationship*

## Z80 Instruction Fetch

The Z80 control signals reflect which machine cycle and T-state is currently being performed by the processor. For the M1 instruction fetch cycle, the control signal sequence is shown in Figure 17-2 [Z80UM].

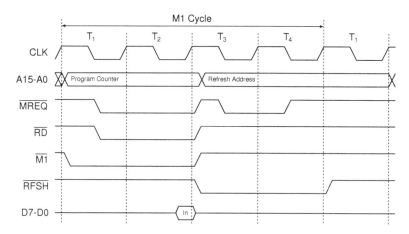

*Figure 17-2: Z80 instruction fetch*

The Z80 begins by placing the address of the required Op Code on the address bus shortly after the start of T-state $T_1$. Half a clock cycle later, midway through $T_1$, the /MREQ signal becomes active indicating that the address on the bus is stable enough to be read by the memory, and may be used directly as a RAM enable signal. At the same time, the /RD signal becomes active, indicating that a byte should be fetched and placed on the data bus. The CPU samples its data bus at the rising clock edge at the start of $T_3$, and then immediately disables both /MREQ and /RD signals. The /M1 signal is active during $T_1$ and $T_2$ to indicate that an instruction fetch is being performed.

During cycles $T_3$ and $T_4$ the instruction is decoded and executed, and while this is taking place no other operation can occur; therefore the CPU buses and control signals are free, and the CPU uses them during this period to perform a memory refresh. A seven bit refresh address is placed on the address bus at the start of $T_3$ and the /RFSH signal goes low, indicating that all dynamic RAM should prepare to perform a refresh read. Halfway through $T_3$, when the address bus has stabilised, /MREQ goes low to indicate that it is safe to use the refresh address. /MREQ is removed at the downward clock transition of $T_4$, followed half a cycle later by /RFSH.

## Z80 Memory Read Or Write

A memory read or write cycle is normally three T-states in length. A read cycle is similar to the instruction fetch cycle in that /MREQ indicates when the address bus is stable, and may be used as a RAM enable. /RD goes low at the same time as /MREQ and indicates that the memory should place the byte to be read on the data bus.

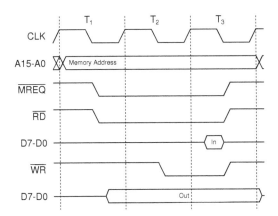

*Figure 17-3: Z80 Memory read and write cycles*

In a write cycle, /MREQ also goes low once the address bus is stable, however the /WR signal becomes active halfway through $T_2$ when the byte on the data bus has stabilised, allowing it to be used directly as a memory read/write select. The /WR signal is removed midway through $T_3$, half a cycle before the values on the address and data bus are removed, so that the CPU complies with the specifications of most memory devices [Z80UM].

## Dynamic RAM Timing Considerations

The DRAM signal timing for the $\mu$PD416 [DS4116] DRAM used in the ZX Spectrum is presented in Figure 17-4.

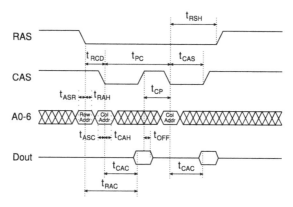

Figure 17-4: 4116 dynamic RAM read and write signal timings

### DRAM Read

As with all DRAM operations, a read begins by presenting the DRAM row address of the location to be read to the DRAM (usually the lower half of the full address), and when stable activating the row address strobe, or /RAS. Once sufficient time has elapsed for the DRAM to register this row address, the column address (usually the upper half of the full address) is presented to the DRAM, followed shortly after by activation of the column address strobe, or /CAS.

For the $\mu$PD416 16K DRAM used in the ZX Spectrum, after /RAS has gone low, the row address must be held for 20ns before the column address is supplied ($t_{RAH}$) and /CAS is taken low ($t_{RCD}$). The column address must be held for 45ns ($t_{CAH}$) following activation of the /CAS.

Once /CAS has gone low, /RAS must continue to be held for at least 100ns ($t_{RSH}$), and /CAS held for a minimum of 100ns ($t_{CAS}$). Data will be available no sooner than 150ns ($t_{150}$) from the start of /RAS or 100ns ($t_{100}$) from the start of /CAS, which ever is the longest.

### DRAM Write

As with the read, a write begins by presenting the DRAM with the row address of the location to be written, and when stable activating /RAS. The write enable signal, /WE, must be enabled before /CAS is activated to put the DRAM into an early-write mode. This prevents the DRAM interpreting the /CAS as a read access and mistakenly placing data on its output. After the row address has been held for 20ns ($t_{RAH}$) the column address can be supplied and allowed to stabilised before /CAS is taken low to store the value. The column address must be held for 45ns ($t_{CAH}$) during the /CAS, which must be low for at least 100ns ($t_{CAS}$), and /RAS held for the same length of time ($t_{RSH}$).

## 16K DRAM CPU Interface

The ZX Spectrum transforms fourteen bits of the 16-bit Z80 address bus into a 7-bit multiplexed DRAM bus with two 74LS157 multiplexer ICs on the ZX Spectrum PCB. The ULA controls which half of the multiplexed CPU address bus is presented to the DRAM in addition to providing the dynamic RAM control signals discussed previously.

The ULA is connected to the DRAM through tri-state outputs so that it can disconnect itself from the DRAM when the Z80 has access. Similarly, the Z80 must be isolated from the DRAM when the ULA has access. Ideally, tri-state buffers would have also been used, but because this increased the number of ICs and PCB space required, directly translating a into higher cost, Altwasser borrowed a technique first used in the ZX80, and isolated the Z80 buses by connecting them to the DRAM via 330 ohm resistors. Since the ULA generates all the lower 16K DRAM control signals, it is able to take the steps required to release the DRAM address bus over to the CPU as soon as it finishes its video byte fetch, and put its own address bus into a high impedance state. This bus isolation is shown in Figure 17-5.

In the ZX Spectrum, the largest unit of memory common to its three memory devices of ROM, 16K RAM and 32K RAM is 16K or 16384 bytes. This range of addresses may be represented by 14 bits ($16384 = 2^{14}$), and thus requires address lines A13–A0 to access all 16384 locations. The ZX Spectrum's full 64K can therefore be split into four 16K blocks, as shown by Table 17-1.

| Address | A15 | A14 | Memory Device |
|---|---|---|---|
| 0 – 16383 | 0 | 0 | 16K ROM |
| 16384 – 32767 | 0 | 1 | 16K RAM |
| 32768 – 49151 | 1 | 0 | 32K RAM – First 16K half |
| 49152 – 65535 | 1 | 1 | 32K RAM – Second 16K half |

Table 17-1: ZX Spectrum memory mapping by 16K block

The two remaining CPU address bus lines, A14 and A15, determine which of the four 16K blocks is being accessed; thus by considering combinations of A14 and A15, the appropriate memory device is selected. Note that whenever A15 is 1, the 32K memory present in 48K machines is selected, regardless of the state of A14. Instead A14 is used by the 32K memory chips to select one of its two internal 16K banks.

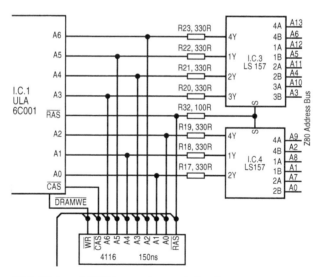

Figure 17-5: ULA and CPU bus and control signal connection to 16K DRAM

## ZX Spectrum ROM Select

Altwasser followed the Zilog specification and used the CPU /MREQ signal as the primary memory enable, decoding the address bus when it becomes active and enabling the ROM if the CPU is addressing a memory location between 0 and 16383. For these locations, address lines A14 and A15 will both be low (Table 17-1).

Like most memory devices, the ROM requires an active low enable signal, and the ULA ROM enable logic is relatively simple:

$$\overline{ROMCS} = A_{14} + A_{15}$$

The schematic for /ROMCS is presented in Figure 17-7 and is subsequently fed to peripheral cell 27, where it is conditioned as a TTL output and passed to pin 34.

## CPU RAS Generation

When the processor accesses the 16K DRAM by placing an address between address 16384 and 32767 on its bus, /MREQ will go low with A15 low and A14 high, as given in Table 17-1. These signals are decoded by the ULA into an internal signal RAM$_{16}$, which it uses to activate the generation of the other DRAM signals necessary to provide the CPU with access to the 16K DRAM:

$$RAM_{16} = \overline{MREQ} + \overline{A_{14}} + A_{15}$$

The engineers at Zilog designed the Z80 so that /MREQ could be used directly as a DRAM RAS, and places the address to be accessed on its bus half a clock cycle before /MREQ, allowing 142ns for the address to stabilise, given a 3.5MHz clock. Since RAM$_{16}$ is synchronous with /MREQ for addresses between 16384 and 32767, the ULA uses it as the CPU /RAS for the 16K DRAM.

The ULA has a single /RAS pin through which it sends a combined video and CPU RAS to the DRAM, and this combined signal is created by NOR gating RAM$_{16}$ and the video RAS signal (described in the section called *RAS Generation* in Chapter 13, *Video Memory Access*):

$$\overline{RAS} = \overline{RAM_{16} + VidRAS}$$

This signal is routed to peripheral cell 26, where it is conditioned as a TTL output and connected to pin 35. See Chapter 22, *Signal Interfacing*.

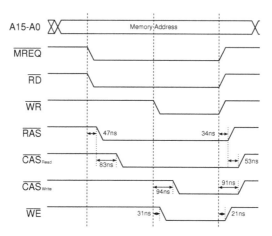

*Figure 17-6: CPU, RAS and CAS related timing*

## CPU CAS Generation

Zilog engineered the Z80 so that the /RD and /WR signals can drive the generation of the DRAM column address strobe with little additional logic. Generation of the CPUCAS signal is controlled by the internal $RAM_{16}$ signal, which is active when the CPU is addressing a location between 16384 and 32767. RAM16 can therefore be combined with /RD and /WR to generate a CAS signal whenever the CPU performs a read or write to the 16K DRAM.

$$CPUCAS_{simple} = \overline{\overline{(RD + WR)} + \overline{RAM_{16}}}$$

To comply with the specifications of the $\mu$PD416 150ns dynamic RAM [DS4116], a period of $t_{RCD}$ must pass between the activation of RAS and the activation of CAS. When the CPU is performing a write operation, /WR goes low one clock cycle after /MREQ, and as a result CPUCAS defined by the equation above will go low 285ns after CPURAS, given the 3.5MHz CPU clock. As this exceeds the minimum $t_{RCD}$ of 20ns required by the DRAM, no additional delay needs to be added. However, when performing a read, the CPU /RD line goes low at the same time as /MREQ, which results in

CPUCAS and CPURAS also going low at the same time; therefore the generation of CPUCAS must include a delay of at least $t_{RCD}$ (20ns) to avoid this situation.

The ULA implementation of CPUCAS incorporates a number of dummy logic gates that introduce some signal delay. The following equations and description assume an 10ns gate propagation delay, and should be referred to Figure 17-7.

Typically with DRAM control circuits, the generated CAS signal is delayed twice. First to produce a multiplexer select signal (MUXSEL) that switches the bus over to the column address some time after /RAS has gone low, and second to allow the column address to settle before the CAS is activated.

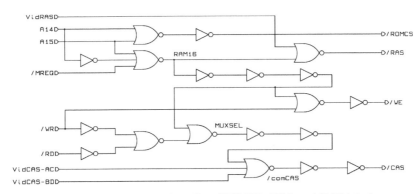

*Figure 17-7: CPU memory decoding, /ROMCS, /RAS and /CAS interfaces*

First RAM$_{16}$ is delayed and inverted by three gates before it is NOR-gated with a combined /RDWR signal, creating the MUXSEL signal that is delayed with respect to RAM16, and therefore CPURAS by approximately 30ns:

$$MUXSEL = \overline{\overline{(RD + WR)} + \overline{RAM_{16(-30)}}}$$

It should be noted that this first delay only takes effect during a read operation, where /RD goes low with RAM16. During a write, /WR goes low 285ns after RAM16 and overrides its delay; however 285ns is more than enough time between /RAS and MUXSEL for the DRAM to register the row address.

The Interestingly this MUXSEL signal is not used by the ULA, presumably because of the limited number of IC pins available, and the multiplexers on the ZX Spectrum PCB instead make use of the /RAS signal as their select.

MUXSEL is further delayed by two inverters, producing the CPUCAS signal which is combined with the two video CAS signals VidCAS$_{AC}$ and VidCAS$_{BD}$, to create a combined /comCAS signal (this combination is also shown in the section called *CAS Generation* in Chapter 13, *Video Memory Access*):

$$\overline{comCAS} = \overline{CPUCAS_{-30} + VidCAS_{AC} + VidCAS_{BD}}$$

This three input NOR gate imparts some delay to the signal as well as the two inverters that follow it, delaying the /compCAS signal by a further 30ns to create the final /CAS:

$$\overline{CAS} = \overline{compCAS_{-30}}$$

The overall delay applied to /CAS with respect to RAM$_{16}$ is approximately 90ns. /RAS has a 10ns delay with respect to the same signal, giving a difference of approximately 80ns between /RAS and /CAS. The /CAS signal is routed to peripheral cell 19, where it is conditioned as a TTL output and connected to pin 1. See the section called *RAS Generation* in Chapter 13, *Video Memory Access*):

Note that the interface circuits contribute to the overall signal delay measured between the ULA pins. For instance, a large part of the delay between /MREQ and /RAS can be attributed to the /MREQ input and the /RAS output delays, adding between 20 and 30ns onto the /MREQ to /RAS timing. Figure 17-6 shows the CPU, RAS and CAS related timings, measured on a 6C001E-7 ULA. Of the 47ns delay measured between /MREQ and /RAS, approximately 20ns of this is due to delays in the /RAS generation circuit shown in Figure 17-7, and the remainder due to interface delay.

## Memory Write Control

In addition to controlling CPU access to the lower 16K of RAM, the ULA must make sure that the CPU write enable signal, /WR, is not applied to the RAM when the CPU is not, or is prevented from, accessing it. This is to guarantee that the RAM does not interpret a video display fetch as being a write operation, whenever the CPU activates its /WR signal.

To do this the CPU /WR signal is fed to the ULA where it is gated internally with /RAM$_{16}$ to produce a new write enable signal, /WE. The delayed RAM16 signal is used merely for convenience in the interconnect routing, and has no effect since CPU /WR goes low after the delay has past. This /WE signal goes low with the CPU /WR line when the CPU has access to the lower 16K of RAM, and is held high at all other times.

$$\overline{WE} = \overline{RAM_{16(-30)}} + \overline{WR}$$

---

1. Upgrading a 16K issue 1 machine to 48K required a daughter board to be plugged into the main circuit board, a 16K issue 2 machine required eight memory and four logic ICs to be plugged into the PCB sockets provided.

# Chapter 18
# CPU Clock and Contention

The ZX Spectrum Z80 CPU is driven by a 3.5MHz clock, derived by the ULA from its internal counter bit, C0. This frequency was chosen as it was easily obtained by dividing the 14MHz ULA master clock by four, and allowed the 4MHz Z80A CPU to run at almost full speed.

The Z80 requires an accurate, square clock to control and sequence its internal state machine. A non-square clock causes its various circuits to each switch at unspecified points in the clock cycle, leading to erratic behaviour. Consequently, to guarantee that the Z80 switches accurately, the clock signal generated by the ULA is designed to receive further buffering and shaping before being fed to the Z80; therefore an inverted clock is produced which is subsequently amplified and re-inverted by a transistor buffer on the ZX Spectrum circuit board, producing a clean square wave signal that rises and falls rapidly between zero and five volts.

## Memory Contention

As discussed in Chapter 13, *Video Memory Access* and Chapter 17, *CPU Memory Access*, both the Z80 and video controller require access to the lower 16K of RAM. The RAM can only perform a single operation at a time, and as the display cannot be interrupted, the video controller takes precedence. This means that to avoid contention the Z80 must be prevented from accessing the memory while the video controller is doing so.

The first step in determining whether memory contention will occur is to identify a hazardous address appearing on the Z80 address bus while the video controller is accessing the lower 16K RAM. Any Z80 address falling between 0x4000 and 0x7FFF (16384 and 32767 decimal), inclusive, is an access to the video RAM and may conflict with the video controller, see Chapter 17, *CPU Memory Access*. On detecting an address in this range, the video controller must pause the Z80 until it has finished its video read operation.

There are two methods for pausing the Z80 described by the data sheet [Z80UM], but Altwasser found them to be unsuitable when he considered the additional signals required to control them and the precise timing demands of the video controller. Altwasser was forced, therefore, to invent a third method which could be synchronised to the video controller and which did not require any additional control signals and ULA pins to interface to the Z80. All three methods are described below.

### Z80 WAIT states

The Z80 provides a special control pin, /WAIT, to assist with interfacing to slow memory devices. The Z80 samples this pin at the falling edge of the clock during the second T-state of an instruction cycle, $T_2$. If it is found to be low, the Z80 extends the second T-state by one clock cycle. The /WAIT pin is then sampled again and the process repeated until /WAIT has returned to a high condition. Once the wait period is over, the Z80 completes the second T-state and proceeds with the third, where data is read from or written to the memory. See Figure 18-1.

The lower 16K of RAM can be treated as a slow memory device while the video controller is reading display information, and therefore this mechanism for delaying the Z80 memory access would satisfy the requirements for contention resolution.

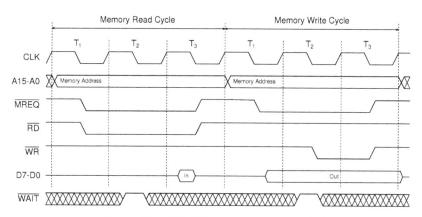

*Figure 18-1: Z80 CPU memory operation timing*

However, closer examination of the Z80 signal timing reveals that the address bus and control signals /MREQ, /RD and /WR, are established during $T_1$, one

T-state before the /WAIT control pin is sampled. This would be an issue for the ULA as it has no internal override of the CPU RAS and CAS generation discussed in Chapter 17, *CPU Memory Access*[1]. As soon as the Z80 activates the /MREQ line with the address bus holding a value between 0x4000 and 0x7FFF, the ULA generates a /RAS signal on behalf of the Z80. This would interfere with the /RAS signal that the ULA video controller may be generating and result in an incorrect display update. Had it not been for the lack of spare pins on the ULA to allow connection of the /WAIT signal, an override of the CPU RAS-CAS production could have been implemented with what little spare logic remains in the ULA, allowing the /WAIT mechanism to be used.

## Z80 Bus Request

The second mechanism provided by the Z80 to allow external devices to pause its operation is call a bus request. The /BUSREQ pin is sampled by the Z80 at the rising edge of the clock at the start of the last T-state of a machine cycle, see Figure 18-2. If it is found to be low, then at the next rising edge of the CPU clock the Z80 puts its address bus, data bus and control signals into a high impedance state, and at the rising edge of each subsequent clock period, /BUSREQ is sampled again. This continues until /BUSREQ returns high, when, at the rising edge of the following clock period, the buses and control signals are returned to normal. This allows external devices to stop the Z80 and take control of its buses for as long as required.

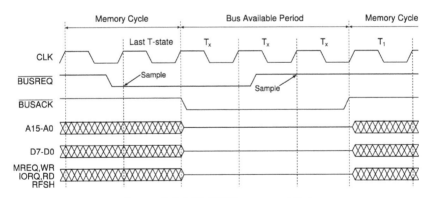

*Figure 18-2: Z80 CPU bus request timing*

Even though this mechanism detaches the Z80 from its buses and control signals, closer inspection of the timing involved reveals that since the /BUSREQ sampling occurs during the last T-state of a machine cycle, this takes place during $T_4$ of an instruction fetch, which is after the memory read has been completed. This means that instruction fetches cannot be interrupted using this method, nor can the activation of the Z80 memory control signals be prevented prior to the bus request being detected and accepted.

This method of temporarily suspending the Z80 would therefore interfere with the video generation in a similar way to the wait state approach described previously, and would again require a dedicated ULA pin to pass the bus request to the Z80, when all available pins were committed to other tasks.

### Clock Interruption

The method of pausing the Z80 that was designed by Altwasser and employed in the ULA, is at once simple, elegant and clever. It does not require a dedicated pin of the ULA or the additional complication of a CPU RAS-CAS override. It works by stopping the clock signal it sends to the Z80 at strategic T-states, whenever it detects a contention condition is about to occur. The technique exploits the timing of the Z80 clock and memory control signals, and demonstrates the in depth understanding the ZX Spectrum designers had of the Z80.

What makes this method of stopping the Z80 possible is that the address of the memory location to be accessed is placed on the address bus half a clock cycle before the memory control signals are activated. See Figure 18-1. The Z80 introduces this delay to ensure that the address on the bus has stabilised before any device attempts to read it. The ULA monitors the address bus and when it detects a CPU address between 16384 and 32767 while it is reading video bytes, it immediately holds the Z80 clock high. Because this happens during the first half of state $T_1$, the downward clock transition does not occur and the Z80 memory control signals remain inactive as a result. Consequently the CPU RAS-CAS signals are not generated by the ULA, and no interference with the video generation occurs.

By holding the clock high at the start of the memory access state $T_1$, the ULA lengthens the T-state until the video controller has completed its memory fetch sequence. The Z80 is unaware that this is happening, as its only source of time reference is the clock. It is during $T_2$ and $T_3$ that a memory write takes place, or $T_3$ alone for a memory read; therefore these T-states are never interrupted, and the ULA allows sufficient time after releasing $T_1$, for T-states $T_2$ and $T_3$ to complete before the video controller requires memory access again. Because the master counter is negative edge triggered, ULA signals derived from it are usually aligned to these negative edges. In contrast, each Z80 clock cycle starts with the positive edge of its clock, which is derived from C0; therefore ULA signals will generally transition halfway through a processor T-state.

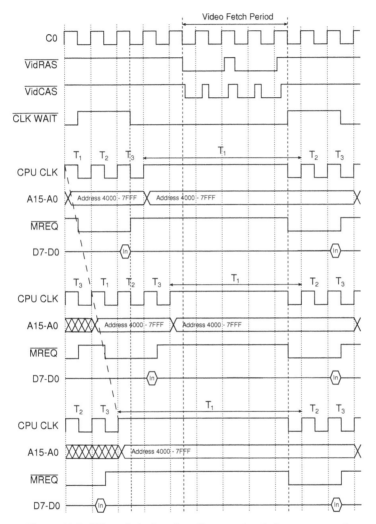

Figure 18-3: Effect of clock wait on three contended memory read operations

The video fetch takes four out of eight C0 cycles to complete a read, spanning five CPU T-states, illustrated by Figure 18-3. All Z80 memory read and write operations complete in three T-states[2] and this was an important consideration when deciding how many T-states for which to hold the Z80 clock during a memory access conflict.

## Clock Wait Generation

As discussed previously, $T_1$ is the only candidate suitable for T-state length-ening because only here is the address bus enabled before any control signals. The two T-states that follow are always allowed to proceed without interrup-tion, and so the period during which $T_1$ is checked and possibly held (clock wait) must begin two T-states before the start of the video fetch. This prevents $T_1$ from executing within two T-states of the start of the video fetch, leaving time for $T_2$ and $T_3$ to complete before the fetch begins. This gives a total wait period of six cycles, as demonstrated in Figure 18-3, which shows a memory access starting at three consecutive cycles of C0 and the effect of the clock wait period on their execution. Again, note that the CPU T-states and signals are offset from the ULA signals by half a C0 cycle.

The memory contention clock wait signal is derived from C2 and C3, as fol-lows:

$$\overline{CLKWAIT} = \overline{C3 + C2}$$

The signal timing is critical here, because the clock wait transitions are aligned with the negative going edge of C0. Any delay of the clock wait signal such that it lags behind C0 will cause Z80 stability problems, since CPUCLK will be allowed to go low before the wait signal is able to pull it back high and hold it there, causing a momentary low processor clock spike.

## T1 Start Detection

During a memory operation the Z80 address bus contains a valid address from the start of $T_1$ through $T_2$ and $T_3{}^3$. The ULA contention controller monitors A14 and A15 of the address bus, and if an access is made to the lower 16K RAM during $T_1$, the clock wait signal is applied to the Z80 clock to hold it high. As the clock wait signal is active only when a video fetch is about to occur, or is in progress, the Z80 operation is suspended only when a contention condition exists.

Determining that T-state $T_1$ has begun can only be achieved by looking for a specific address, as the processor will have only activated the address bus at this point. Preventing contention checking outside of this T-state is achieved as follows. /MREQ goes low halfway through the first T-state, at the downward transition of the clock, and by delaying it until the clock goes high at the end of the T-state, an intermediate signal MREQT23 is produced that goes high for T-states $T_2$ and $T_3$, disabling contention checking.

## I/O Contention

In addition to memory contention between the video controller and the Z80, it is also possible for the ULA and the Z80 to be in contention when the Z80 requests access to the ULA I/O port. See Chapter 19, *Input-Output Devices*. The ULA interfaces to external devices such as the Z80 and memory through a single data bus and set of control signals. If the Z80 requires access to the ULA I/O port while a video fetch is underway, the ULA data bus will be committed to transferring video data and will be unable to satisfy the I/O request.

To avoid a conflict, the ULA holds the Z80 clock high whenever it detects the Z80 attempting to access its I/O port during a video fetch sequence.

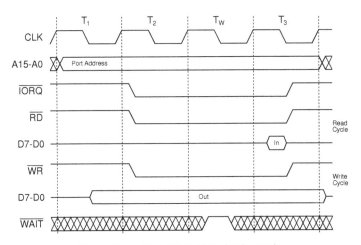

*Figure 18-4: CPU I/O read and write cycles*

The ULA I/O port is officially documented as being at address 0x00FE (254 decimal), however this is not technically accurate as Sinclair machines implemented partial decoding of the address bus for I/O ports to reduce complexity. In fact the ZX Spectrum decodes the single address line A0 which, if low during an I/O request, signifies an access to the ULA I/O port. This means that *any* port address that has A0 low is interpreted as a ULA port access. See

Chapter 19, *Input-Output Devices* for further details of this I/O port address decoding.

In general, memory operations take three T-states to complete, whereas I/O operations take four. The first T-state places the I/O address on the address bus, the second activates the control signals /IORQ and /RD or /WR, the third is a special T-state $T_w$ where the /WAIT pin is sampled, and the fourth, officially labelled $T_3$, is where the I/O operation is performed.

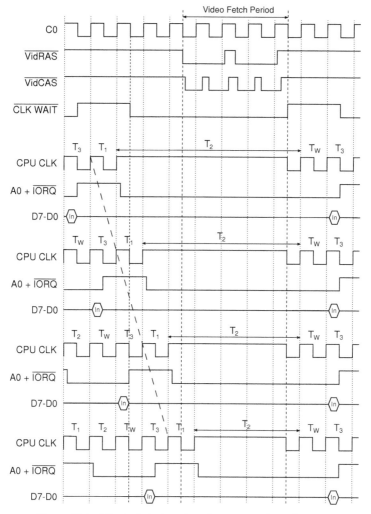

*Figure 18-5: Effect of clock wait on four contended I/O read operations*

The ULA cannot use the address bus alone to determine whether an I/O access is being made by the Z80, as an I/O address is indistinguishable from a memory address during the first T-state of a memory operation Thankfully, unlike the memory control signals, there are no adverse effects from having the I/O control signals active during a video fetch, allowing the ULA to safely wait for /IOREQ to go low at the start of I/O T-state $T_2$ before it decides whether or not to hold the Z80 clock.

I/O T-state $T_2$ is therefore the first detectable T-state prior to the I/O operation being carried out, and hence the T-state that the ULA lengthens if it detects a ULA I/O port access during a video fetch. The remaining two T-states of the machine cycle are allowed to continue uninterrupted, with the I/O operation being performed by the final T-state. This sequence mirrors that of the memory contention model, with the wait period starting two T-states before the video fetch begins. This prevents $T_2$ from executing within two T-states of the video fetch, and allows time for $T_W$ and $T_3$ to complete uninterrupted.

Using the same clock wait signal as memory contention does not give the optimal contention timing for I/O operations, as the clock will be held longer than necessary. Altwasser's efforts to reduce this period is described in the section called *The issue 1 5C102 ULA*, and the timing given in Figure 18-5 is that ultimately implemented by all pre-Amstrad Spectrum models.

### I/O T2 Detection

The start of I/O T-state $T_2$ is indicated by the Z80 signal /IORQ going low, and ends at the next positive edge of the CPU clock. The ULA uses the /IOREQ signal (the result of OR gating /IORQ with A0) to indicate that the $T_2$ of a ULA port access has begun, and /IOREQ delayed until the next rising edge of the CPU clock to indicate that $T_2$ has finished and states $T_W$ and $T_3$ are in progress.

This delayed signal, named /IOREQTW3, is active only during the final two T-states of a ULA port access, during which time the Z80 has uninterrupted access to the ULA I/O port. It is therefore used by the ULA to activate its internal I/O port handling circuitry, see Chapter 19, *Input-Output Devices* for further details.

Due to its reliance on /IOREQ to differentiate between memory and I/O contention, and because the address bus becomes active at the start of $T_1$, the ULA will enter a contended mode at the start of $T_1$ when an even (A0 low) I/O address between 0x4000 and 0x7FFF is accessed. When an odd (A0 high) I/O address in the same range is accessed, I/O contention is applied for the entire

I/O instruction cycle, as /IOREQ (being the logical OR of A0 and /IORQ) will never go low to trigger its cancellation at the end of $T_2$.

## Circuit Description

A number of signals are brought together to control the Z80 clock during times of memory and I/O contention between the video controller and the Z80:

1. *C0*: The 3.5MHz clock that forms the basis of the Z80 clock.

2. *Border*: This signal disables contention detection while the video controller is displaying the border, and therefore not performing a video update.

3. *A14 and A15*: These two Z80 address bus signals indicate, among other things, which 16K block of memory the Z80 wants to access. See Chapter 8, *The Memory Map*). The address bus would normally be considered along with /MREQ when decoding memory accesses.

4. */MREQ*: This signal indicates that the Z80 is performing a memory operation.

5. */IOREQ*: This signal indicates that the Z80 is accessing the ULA I/O port, being the result of OR gating /IORQ with A0.

6. *C2 and C3*: When combined, these two master counter signals indicate the periods during a scan line that the video controller might require access to the video RAM. They ultimately determine the length of the wait signal that holds the Z80 clock high when the control circuit detects a contention condition.

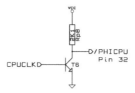

*Figure 18-6: CPU clock output driver at peripheral cell 28, pin 32*

To produce the Z80 clock the ZX Spectrum takes /C0, oscillating at 3.5MHz, inverts it and gates it with a clock control signal produced by the contention controller. The resulting clock signal is inverted and passed to a peripheral cell where it is amplified to a peak level of 5v by a transistor in a collector-follower configuration. This transistor also inverts the signal back to a negative phase before it is passed out of the chip through ULA pin 32 (Figure 18-6).

The contention controller is divided into two parts: one that detects and controls memory contention, the other that detects and controls I/O contention. These two processes combine to form the clock control signal that gates C0 to produce the Z80 clock.

The contention controller has been the subject of several revisions during the course of the ZX Spectrum's manufacture, as early ULA versions contained subtle design errors that affected the reliability of the machine. The operation of the contention controller in each ULA version is discussed below.

### The issue 1 5C102 ULA

The issue 1 memory contention handler considers A14 and A15, looking for an address between 0x4000 and 0x7FFF, and delays /MREQ until the end of the T-state by passing it through a gated D transparent latch clocked by the CPU clock. During Z80 memory cycles this delayed signal, designated MREQT23, is high for T-states $T_2$ and $T_3$, but low for $T_1$. The contention circuit detects that the Z80 is executing the first T-state of a contended memory cycle by waiting for A14 to go high, with A15 and MREQT23 low.

The contention handler combines C2 and C3 to produce the clock wait signal /MWAIT, which defines the period during which contention could occur, discussed in the section called *Clock Wait Generation*. While this signal is high, the clock control signal from the contention handler is forced low and switched off. Therefore, whenever a contended T-state $T_2$ is detected, the output of the contention handler is governed by /MWAIT.

Also considered is the Border signal, which indicates when the video controller does not require access to the video memory and contention checking can be disabled. When high, the Border signal forces the clock control signal low, allowing C0 to pass uninterrupted through the output NOR gate as CPUCLK.

Lastly /CPUCLK is fed back into the contention circuit to hold it inactive while CPUCLK is low. This avoids glitches in the clock output by only allowing CPUCLK to be held high from an already high state, as discussed in the section called *Clock Wait Generation*.

The I/O contention handler delays changes to /IOREQ until the end of the T-state by passing it through a gated D transparent latch clocked by the CPU clock. This creates the IOREQTW3 signal, which is high during $T_w$ and $T_3$ of an I/O cycle. By bringing together /IOREQ, which goes low during $T_2$, and IOREQTW3, the contention controller detects the start and end of I/O T-state $T_2$, allowing it to activate the clock control as required. Like memory contention this control signal is forced low, and thus deactivated, while Border is high or CPUCLK low.

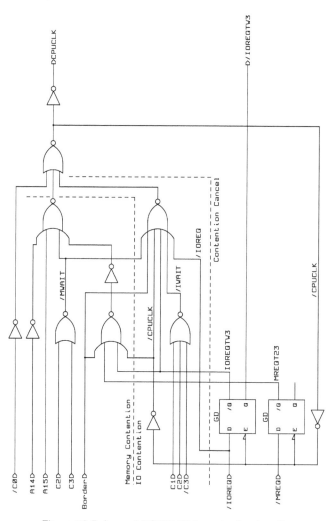

*Figure 18-7: Issue 1 5C102 ULA contention handler*

The 5C102 ULA wait timing for I/O contention is different from that of memory contention, and is given by the following simplified expression:

$$\overline{IOWAIT} = \overline{(C3 + C2)} + \overline{(C2 + C1)}$$

This I/O clock wait signal is active for a total of five C0 cycles in every eight, while the CPU is executing contended ULA I/O operations, one cycle less than for memory contention. It should be noted that the clock wait is released for one cycle following the first two contended periods of C0. This incorrectly allows any I/O instruction being held to continue execution from that point, resulting in a data bus collision between the second and third byte fetches with the I/O transfer of $T_3$, see Figure 18-8.

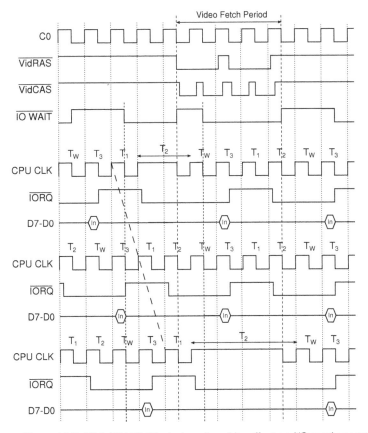

Figure 18-8: ULA 5C102 I/O clock wait and its effect on I/O read operations

Because the ULA has direct access to the memory, and the CPU is buffered via resistors, this conflict does not interfere with the display update but instead corrupts the I/O port value being read or written. The effect is to make I/O operations such as reading the keyboard erratic, as their success depends on whether the ULA is updating the display.

This fault passed undetected initially because the ZX Spectrum operating system reads the keyboard via an interrupt service routine, executed while the television electron beam is returning to the top of the screen, and therefore during a period when contention does not occur. As soon as software began to be written in machine code that read the keyboard I/O port directly, it was discovered that the read was unreliable and suffered from contention problems.

Analysis showed that the I/O clock wait signal should not be released for the third C0 cycle, and that the combination of C1 and C2 were to blame. Sinclair was faced with the prospect of discarding the first batch of ULA chips, at a huge financial loss, until it realised that by manipulating the signals passed to the ULA from the CPU, I/O contention could be made to look like memory contention and would therefore be subject to the more aggressive memory clock wait timing.

The signal manipulation took the form of a 14 pin NAND gate integrated circuit placed upside down on the circuit board and connected between address lines A14 and A15 as they ran from the Z80 to the ULA, as presented in Figure 18-9. This modification forced A14 high and A15 low whenever /IORQ went low, creating a condition that was recognised by the ULA as a contended memory access. This modification subsequently became known as the "dead cockroach", as pictured in Figure 18-10.

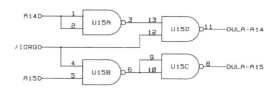

*Figure 18-9: Cockroach modification added to issue 1 ULA machines*

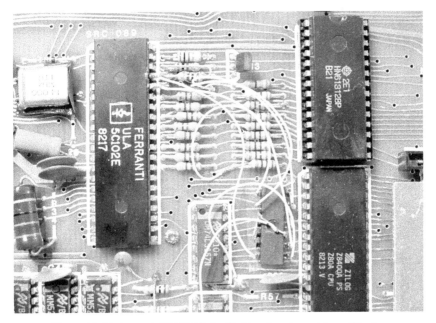

*Figure 18-10: Issue 1 5C102 ULA with "dead cockroach" modification*

The intention behind the I/O clock wait timing is likely to have been to reduce the contention applied to I/O instructions to a minimum. Because an I/O instructions data transfer occurs during $T_3$ and the I/O control signals do not interfere with display byte fetches, data bus collisions can be avoided by making sure that $T_3$ alone is prevented from executing while video byte fetches are occurring. This is in contrast to contended memory access where T-states $T_1$, $T_2$ or $T_3$ must not execute while video byte fetches are occurring.

This theoretically reduces the I/O contention window to four T-states instead of the six required during memory contention. However, this would allow $T_3$ to execute immediately after a byte fetch sequence has completed, when the video RAM will be releasing the data bus. Because of this, an extra T-state of contention is necessary to allow the data bus to stabilise before $T_3$ performs its data transfer. This ideal I/O contention timing is presented in Figure 18-11, and defined as:

$$\overline{IOWAIT_{IDEAL}} = \overline{(C3 + C2)} + \overline{(\overline{C3} + \overline{C2} + \overline{C1})}$$

It is mere speculation as to whether this was the intended I/O contention timing of the 5C102 ULA, but the suggestion is given weight by the fact that the observed behaviour would be demonstrated if there was an error in the matrix cell interconnection of /C2 and /C1 for the above logic, where the non-inverted signals C2 and C1 were used by mistake, since

$$\overline{(C3 + C2)} + \overline{(\overline{C3} + C2 + C1)} \equiv \overline{(C3 + C2)} + \overline{(C2 + C1)}$$

This gives the only credible explanation for the presence of C1 + C2 in the I/O clock wait circuit, and the reason /C3 is represented in Figure 18-7.

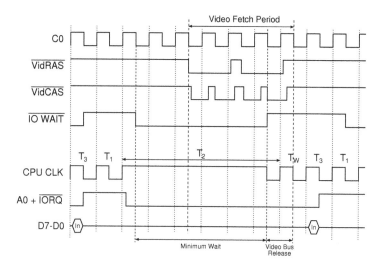

Figure 18-11: The ideal I/O clock wait timing

Sinclair's subsequent "dead cockroach" fix overrides the I/O clock wait signal by forcing activation of the memory clock wait signal when /IOREQ goes low. This established the longer contention timing that would have to be reproduced by all future ULA versions if software were to behave consistently on all machines.

The two contention handlers normally operate independently, but can interact with each other under certain circumstances. Because the memory contention handler interprets an address in the range of 0x4000 to 0x7FFF while /MREQ is high as the start of a contended memory access, accessing an I/O port in this range will activate the memory contention handler during $T_1$ and the I/O contention handler during $T_2$. To ensure that under these conditions all contention checking is disabled at the end of I/O T-state $T_2$, IOREQTW3 is also considered by the memory contention handler.

It is worth noting that the memory contention handler shown in Figure 18-7 uses a seven input NOR gate to produce the clock control signal, the implementation of which is split across two multi-input NOR gates connected together via an inverter. Normally multi-input NOR gates of any size are constructed by connecting in parallel several NOR gates that are in close proximity. However, in this case the NOR gate combining the Border, CPUCLK, MREQT23 and IOREQTW3 signals is sufficiently far away from the output NOR gate for this to be impractical.

**The issue 2 5C112 ULA**

The second issue ULA was produced from circa August 1982, removing the I/O contention problem of the 5C102 ULA. This new ULA essentially had the "dead cockroach" modification incorporated internally, so that whenever /IOREQ went low it overrode address lines A14 and A15, activating the memory contention handler, as shown in Figure 18-12.

Sinclair did not attempt to fix the earlier fault by correctly implementing the intended shorter I/O contention period, as this would have constituted a measurable timing change from the modified 5C102E machines, causing software to behave differently.

Consequently the I/O specific contention handler became effectively redundant as the memory contention handler assumed responsibility for both memory and I/O contention. The I/O contention circuit remained in place however, as the absolute minimum was changed of the ULA interconnection layer to produce the new ULA as quickly and cheaply as possible. See Figure 18-12.

Shortly after the issue 2 ULA went into production, a further fault was discovered with both the issue 1 and 2 ULAs that prevented external peripherals such as the ZX Printer, and later the interface 1 and Microdrive, from operating correctly.

The problem was that the ULA I/O port address was not considered with /IORQ when deciding whether I/O contention was about to occur. This resulted in I/O contention being applied when *any* I/O port was accessed, pausing the processor.

Luckily a minimum of address line A0 needed to be considered, as discussed in the section called *Decoding the I/O Port* in Chapter 19, *Input-Output Devices*, and thanks to resistor R27 having being included in the /IORQ signal passed to the ULA from the processor, a simple OR gate was created by the addition of a transistor soldered across the Z80 processor, shown in Figure 18-14. This pulled /IOREQ high whenever A0 was high, overriding /IORQ and disabling I/O contention checking. See Figure 18-13.

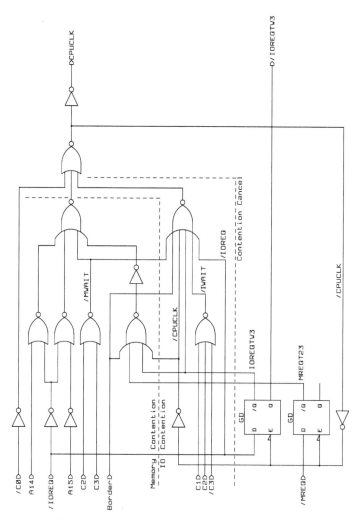

*Figure 18-12: Issue 2 5C112 ULA contention handler*

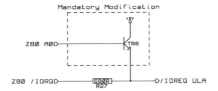

*Figure 18-13: Transistor and resistor OR gate of /IORQ and A0*

This transistor became a standard component of the ZX Spectrum circuit board, and the problem was not corrected in subsequent versions of the ULA.

*Figure 18-14: Issue 2 ZX Spectrum with "spider" transistor modification*

**The issue 3 6C001 ULA**

The third issue ULA, produced circa May 1983, incorporates a number of enhancements to the design implemented by the previous versions, and was realised by a more advanced ULA product from Ferranti. Therefore the ULA interconnection layer had to be redrawn for the new ULA geometry, allowing design changes to be incorporated at the same time.

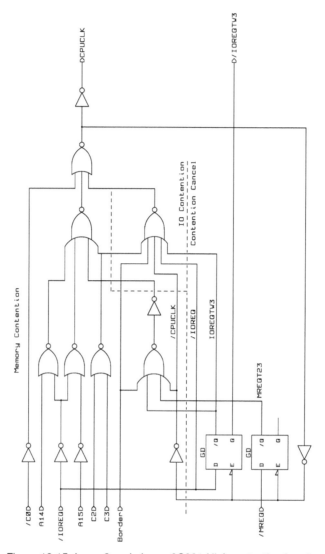

*Figure 18-15: Issue 3 and above, 6C001 ULA contention handler*

One of these design changes was to remove from the I/O contention handler the faulty I/O clock wait signal (/IWAIT) and the NOR gate that produced it, so that the handler used /MWAIT alone and exhibited the same contention timing as the previous ULA versions that processed all contention through the memory handler. Interestingly, the "dead cockroach" modification was also left in place, allowing the memory contention handler to continue detecting memory and I/O contention, the result being two independent circuits taking identical control of the CPU clock under I/O contention conditions.

This double edged fix may have come about as a result of the 6C001 ULA interconnection layer being drawn straight from design schematics incorporating both historical and new modifications. These schematics may have earlier been revised to reflect the removal of the incorrect /IWAIT signal at the time the fault was discovered with the issue 1 ULA, and subsequently modified to show the internal "dead cockroach" modification implemented in the issue 2 ULA.

1. The ZX Spectrum 16K, 48K and ZX Spectrum 128 employ the same simple bus coupling and RAS-CAS generation, making wait cycles via the /WAIT control pin unsuitable. The redesigned Amstrad era +2A/+3 machines on the other hand have more complicated bus interfacing and do in fact use /WAIT to control contention between the Z80 and the ASIC that replaces the ULA in those models. For this reason, the contention pattern of the +2A/+3 machines differs from the earlier Sinclair designed machines.

2. The instruction Op Code fetch, M1, has an additional T-state $T_4$ which performs a memory refresh, but is not itself involved in the memory read or write activity.

3. A Z80 instruction fetch ends and the memory address is removed from the bus at the beginning of $T_3$, shortly after the positive clock edge. Other memory read and write operations complete at the end of $T_3$

# Chapter 19
# Input-Output Devices

In addition to the television display, the ZX Spectrum supports a number of other input and output devices:

1. A 40 key keyboard.
2. A border colour register.
3. Cassette recorder input.
4. Cassette recorder output.
5. An internal speaker.

As discussed in Chapter 3, *The Standard Microcomputer* the Z80 processor used in the ZX Spectrum supports both memory and I/O mapped input and output. Which mechanism should be used depends on the instruction set required, and the amount of system memory. Since the Spectrum's entire 64K address space may be occupied by memory, I/O mapped input–output is used, taking advantage of the Z80's dedicated I/O control signals and instructions.

## The Keyboard

The ZX Spectrum keyboard is implemented as an array of 40 switches. To interface them individually to the processors 8-bit data bus, five input ports would be required, exceeding the pin count of the ULA and remaining free matrix cells.

To reduce the interface demand of keyboards, manufacturers usually multiplex them by arranging the keys in an electrical grid of rows and columns, so that a keypress connects a row and column together. To check for a keypress, software would first select a keyboard row by writing to an output port and second, test the keyboard columns by reading from an input port. The difficulty lies in the awkward shape of a computer keyboard, being wider than the eight columns a single input port can support. It also requires a dedicated row select output port.

To eliminate these problems the ZX Spectrum ULA performs a clever but simple multiplexing trick and manages to handle all keys through a single input port. This reduces the complexity of the interface by a factor of eight, a huge saving when the ULA space was so tight.

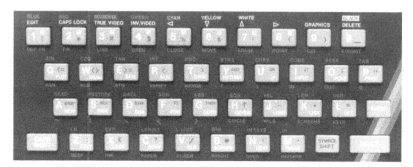

*Figure 19-1: The ZX Spectrum keyboard*

Internally the ULA divides each row of the ZX Spectrum's $4 \times 10$ key keyboard in half, creating eight half-rows of five keys each. These five key columns are assigned to an input port.

To be able to select which of the eight half-rows will be read by the input port without requiring a dedicated row select port, the address of the keyboard input port is chosen to be less than 256, so that it can be decoded by considering address lines A7–0 alone. This leaves the eight upper address lines A15–8 free, and these are used as the half-row selectors. The I/O port address assigned to the keyboard is FE hexadecimal (0xFE), or 254 decimal.

Reading a keyboard half-row therefore depends on the state of the upper address lines, and so the processor sees the keyboard interface as consisting of eight input ports. However, since the I/O address is partially decoded by the ULA and ignores the upper address lines, all eight addresses in fact resolve to a single I/O port.

The advantage of this approach is that multiple keyboard rows may be checked at the same time by enabling more than one of the upper address lines. Under these conditions it is impossible for the processor to differentiate between selected half-rows when a keypress occurs, only the affected column positions is known. Therefore the entire keyboard may be checked for a keypress with one read of the keyboard port, after which individual rows can be examined to discover which key or keys they were.

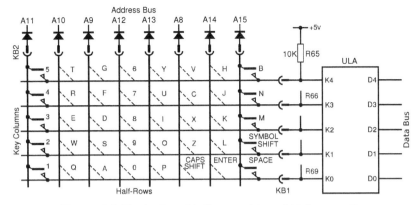

*Figure 19-2: The keyboard half-row matrix and ULA connection*

**Keyboard Input Port**

While no keys are being pressed, the signals of the input port are disconnected from any active circuit and float at an undetermined voltage between 0 and 5v, causing phantom keypresses to be detected. To prevent this, the inputs use pull-up resistors to hold them at logic 1 until a key is pressed, when the corresponding input signal will be pulled down to logic 0 by the address bus. Therefore, setting an upper address line low selects the corresponding half-row.

| Half-Row | High Byte | Address Line | A15 – 8 |
|---|---|---|---|
| B – SPACE | 0x7F | A15 | 0 1 1 1 1 1 1 1 |
| H – ENTER | 0xBF | A14 | 1 0 1 1 1 1 1 1 |
| Y – P | 0xDF | A13 | 1 1 0 1 1 1 1 1 |
| 6 – 0 | 0xEF | A12 | 1 1 1 0 1 1 1 1 |
| 1 – 5 | 0xF7 | A11 | 1 1 1 1 0 1 1 1 |
| Q – T | 0xFB | A10 | 1 1 1 1 1 0 1 1 |
| A – G | 0xFD | A9 | 1 1 1 1 1 1 0 1 |
| CAPS – V | 0xFE | A8 | 1 1 1 1 1 1 1 0 |

*Table 19-1: Keyboard I/O port address high byte values*

Table 19-1 gives the upper address bus values and address line states for each of the eight keyboard half-rows. A row is read from software by setting the low byte of the port address as 0xFE, and the high byte as defined in the table. Multiple rows are read by logically ANDing high bytes together.

Figure 19-2 shows the physical keyboard connection to the input port and address bus. The diodes at the address bus connection prevent multiple simultaneous keypresses in one column from shorting out the processor side of the bus.

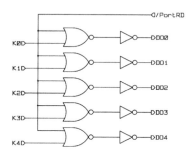

*Figure 19-3: The keyboard input port*

The I/O input port shown in Figure 19-3 consists of a gated buffer that connects the keyboard inputs K4–0 to the data bus D4–0 when the processor makes an I/O read from port 0xFE. The buffer is extremely simple, being made up of just five OR gates. When the I/O port is not being read, the OR gates send logic 1 to the data bus outputs, putting them into a high impedance state and disconnecting them from the bus (see Chapter 22, *Signal Interfacing*). When reading from the keyboard port, the data bus outputs are set low where a key is pressed or high impedance otherwise. These values are registered by the processor as low or high due to the data bus pull-up resistors.

## The Border Colour Register

The television displays the combined outputs of two separate sources. First, the information stored in the video display RAM, and second, a border whose colour is specified by the border colour register.

The border colour register is not implemented as a RAM location, due to the complexity of fetching a byte from memory and the frequency that it will be required. Instead, the register is implemented as a latched three bit I/O output port that directly feeds the display controller. Thus, the border colour value is always available and may be changed at any time with a simple write to the I/O port.

Port address 0xFE has so far only been used for input, and as I/O ports can be both input and output, the same address is used for the ULA output port. This avoids introducing additional port address decoding.

Output I/O ports generally consist of simple gated latches, and the ZX Spectrum is no exception, constructing the border colour register from three gated D transparent latches (Figure 19-4). The register feeds the colour output circuit of the video generator, as discussed in the section called *Border Generation* in Chapter 12, *Generating The Display*.

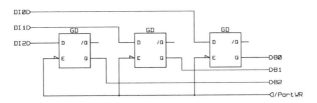

*Figure 19-4: The border colour register*

## Cassette Recorder Input

The linear nature of a cassette tape means that its use as a data storage medium is restricted to serial data streams. The ZX Spectrum ULA must therefore provide a serial interface to send and receive data (see Chapter 3, *The Standard Microcomputer*), and an analogue interface to allow connection to a cassette recorder. The analogue interface is presented in Chapter 20, *Cassette Storage and Sound*.

A hardware serial interface would typically consist of parallel to serial and serial to parallel shift registers, which are complicated and take up a minimum of 40 matrix cells each. Space constraints would not permit this, so the conversion between parallel and serial data is instead carried out in software, interfaced through single bit input and output ports.

*Figure 19-5: The cassette input port*

The cassette input port is implemented in exactly the same way as the keyboard input port; a gated buffer that connects the logic output from the analogue cassette input interface to the data bus, whenever the processor reads from the I/O port. See Figure 19-5.

As only a single bit of an input port is required to implement the cassette input interface, the ULA makes use of one of the free input bits of the keyboard I/O port, in this case D6.

D5 is not used, even though it is the logical choice and its connection pin is closer than D6 to the cassette input buffer (labeled *EAR D6* in Figure A-1). Most of peripheral cell 31, which provides the physical connection to D5 through pin 29, is used for the analogue cassette interface, so the actual interface logic for D5 is provided by peripheral cell 28 and is shared with the CPU clock output PHICLK. D6 is thus the nearest available free data bus output.

Data is stored on cassette as a sequence of tones. These are converted into a sequence of logic pulses by the analogue interface, the frequency of which depends on the tone. The software serial interface samples the I/O port and counts the pulses being received to determine their frequency, and decides whether they represents a binary 0 or 1. Every eight binary digits received are assembled into a byte and stored in RAM.

Conventionally, a cassette recorder output is labeled EAR, and so the ZX Spectrum cassette input port is given the same designation.

## Cassette Recorder Output

Cassette recorder output is similar to input in that it requires a serial interface to convert the parallel data to be written to cassette into a serial stream, and an analogue interface to provide the cassette recorder connection.

The serial interface is once again implemented in software, requiring only a single output port bit through which to send the data stream to the analogue cassette interface.

As with the cassette input port, further use of an existing port is made by allocating the first free bit of output port 0xFE to the cassette recorder output, in this case D3, following the border colour bits. A simple gated D transparent latch stores the value output to this bit and feeds the cassette recorder analogue interface, described in Chapter 20, *Cassette Storage and Sound*.

To store a set of bytes on cassette, the software serial interface takes each bit of each byte in turn, and outputs a sequence of logic pulses to the I/O port, the frequency of which depends on whether the bit was a 0 or 1. The analogue cassette interface converts this pulse sequence into a tone of the same frequency and feeds this to the cassette recorder input.

Conventionally, a cassette recorder input is labeled MIC, and so the ZX Spectrum analogue cassette output port is given the same designation.

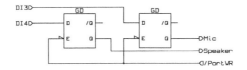

*Figure 19-6: The cassette and speaker output port*

## Speaker Output

The ZX Spectrum does not incorporate a dedicated sound generation device, but instead makes use of a software oscillator that alters the logic state of a single bit output port at varying frequencies. This is analogous to the cassette recorder output mechanism.

The output port is connected to a speaker whose diaphragm moves in or out depending on the state of the port, creating an audible tone at the same frequency.

As with the cassette output port, the next free bit of output port 0xFE is assigned to the speaker output, here being D4, following the cassette output bit. The logic state written to this port bit is latched by a gated D transparent before being fed to the speaker analogue output driver, see Chapter 20, *Cassette Storage and Sound*

## Decoding the I/O Port

As discussed in the section called *The Keyboard*, the ZX Spectrum ULA implements partial decoding of its I/O port to allow the eight most significant address lines to be used as keyboard row selects. Port address 0xFE is represented by 1 1 1 1 1 1 1 0 appearing on address bus lines A7–0, and the ULA takes the partial decoding of this address to the extreme by considering only address line A0.

This partial address decoding scheme was inherited from the ZX80, which was built from SSI TTL chips. Every additional gate increased the risk of requiring an additional integrated circuit, which would raise the component and PCB costs. As this simplified addressing was found to be acceptable, Sinclair's engineers carried it forward in subsequent products.

The consequence of this is that any port address that has A0 low, or put another way, any even numbered address, will select the ULA I/O port. Therefore to avoid accidentally selecting the I/O port of a peripheral connected to the ZX Spectrum, all software must ensure that port 0xFE is addressed so that A7–1 will be at logic 1.

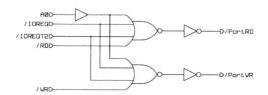

*Figure 19-7: The ULA I/O port decoding of address 0xFE*

The I/O port decoder generates two internal signals, PortRD and PortWR, which are used to enable the input I/O port data bus buffers and output data latches respectively.

The circuit considers A0 and /IOREQ in its decoding of I/O port requests, along with signal /IOREQTW3 generated by the contention handler. /IOREQTW3 is always high during I/O operation T-states T1 and T2, and its inclusion here prevents the ULA I/O port from being activated during T2, when contention may occur. See the section called *I/O T2 Detection* in Chapter 18, *CPU Clock and Contention* for a complete description of how contention affects I/O port access.

## Summary of I/O Port Bit Designations

|  | 7 | 6 | 5 | 4 | 3 | 2 | 1 | 0 |
|---|---|---|---|---|---|---|---|---|
| Input | - | EAR | - | Keyboard | | | | |
| Output | - | - | - | Speaker | MIC | Border | | |

*Table 19-2: I/O port bit map*

# Chapter 20
# Cassette Storage and Sound

In addition to the television output, the ZX Spectrum ULA provides a bi-directional cassette data interface and a single channel audio speaker driver. Both are implemented as software driven serial interfaces, discussed in Chapter 19, *Input-Output Devices*, with analogue input/output stages.

The distinguishing design feature of both the cassette and loudspeaker interface is that they share a single ULA pin. The Sinclair patent *Computer input/output circuit* [ALTWASSERIO] states that "A difficulty is that while these facilities can easily be provided, a substantial number of input/output terminals are usually required. It is an object of the present invention to provide a circuit which enables data and/or programmes to be transferred between a computer and ancillary equipment using a single input/output terminal. Preferably, the circuit is arranged to provide a further output signal at the single terminal for driving an audio output device." This clearly highlights the continual design compromise that was necessary to deliver the ZX Spectrum at the lowest possible cost.

As a result some functionality was lost. Neither is it possible to control the loudspeaker at the same time as reading or writing data to the cassette recorder, nor is it possible to read data from one cassette recorder while writing to another. That said, the software required to perform these activities simultaneously would have been prohibitively complicated to design and execute with an efficient data transfer rate, therefore the compromise is perfectly acceptable.

Altwasser's technical solution is ingenious, and divides the possible voltages output by the I/O pin into two ranges, one that controls the internal loudspeaker, the other that is routed to the cassette output socket. The input interface detects voltage changes at the I/O pin, to which the cassette input socket is also connected, and decides whether the voltage represents a binary one or zero.

## Cassette and Speaker Output

The ZX Spectrum circuit board takes the analogue output from ULA pin 28 and attenuates it with a low pass filter before feeding it to the output socket for the cassette recorder, MIC. Any alternating signal appearing at the ULA I/O pin will therefore be present at the ZX Spectrum MIC socket. The ULA analogue output is also fed to a loudspeaker, having been passed through two series connected diodes. As a diode has a forward voltage drop of approximately 0.7v, the ULA output must exceed 1.4v for the signal to register at the loudspeaker. Thus, alternating voltages below 1.4v appear only at the MIC socket, voltages above 1.4v activate the loudspeaker as well as producing a louder signal from the MIC socket (Figure 20-1). It is by generating voltages within these two ranges that the ZX Spectrum is able to provide both cassette and loudspeaker output from a single analogue source.

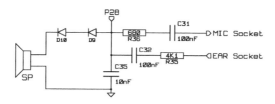

*Figure 20-1: Simplified ZX Spectrum PCB analogue interface*

The cassette and speaker output section of the ULA analogue input/output circuit is shown in Figure 20-2, and consists of a voltage divider with two switchable resistors such that the voltage produced may be set to one of four values. The switching transistors are connected to the two CPU controlled I/O port output bits for the speaker and cassette recorder, described in the section called *Speaker Output* in Chapter 19, *Input-Output Devices* and the section called *Cassette Recorder Output* in Chapter 19, *Input-Output Devices*.

The voltage that is produced at junction A is determined by which combination of switch transistors Q1 and Q3 is active.

Theoretically, the speaker I/O port bit 4 controls transistor Q1 which in turn sets the voltage at the base of Q2, switching resistor R2 in and out of the circuit. When Q1 is off, the base of Q2 is pulled high by R1, switching Q2 on and raising the voltage at junction A of the voltage divider to approximately 4.3v (Vcc minus the base-emitter voltage drop).

When Q1 is on, Q2 shuts off, switching R2 back into the circuit. Assuming Q3 is on and R4 is short-circuited, the voltage divider produces a voltage at junction A of:

$$V_A = \frac{5}{(8100 + 500)} \times 500 = 0.29v$$

The cassette recorder serial output is controlled by I/O port bit 3. When active, this turns transistor Q3 off, switching resistor R4 into the circuit and raising the voltage at junction A. Normally this occurs when the speaker I/O port is inactive, and therefore R2 will also be in the circuit. The voltage measured at junction A is therefore:

$$V_A = \frac{5}{(8100 + 500 + 1500)} \times (500 + 1500) \approx 1v$$

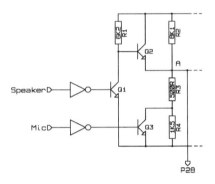

Figure 20-2: ULA 6C001E-7 cassette output and speaker driver

This simplified calculation of node voltages does not take into account the actual voltages applied to the base of Q1 and Q3. In reality, when the MIC port bit is high, the base of Q3 will be at approximately 0.59v (the matrix cell logic low level, as shown in Figure 5-12). This only just keeps the transistor active, with a small amount of current flowing in its collector. Consequently, R4 is not completely switched into the circuit and drops less voltage than would be the case if Q3 were removed, producing a slightly lower voltage at junction A.

When the MIC port bit is low, the base of Q3 will be at the matrix cell logic high level of 0.95v, turning the transistor almost fully on. Current still flows in the collector of Q3 and a small voltage drop appears across R4, raising the voltage at junction A slightly higher than if Q3 were completely on and short circuiting R4.

The speaker control circuit experiences similar residual current flows through Q2 and R2, raising and lowering the voltage at junction A by a small amount.

SPICE analysis of the circuit for an approximate NPN transistor has revealed the node voltages reproduced in Table 20-1, giving the voltage at the base of Q2, the voltage at the collector of Q3 and the voltage at junction A. The actual measured output voltages for the 5C and 6C ULA are given in Table 20-2, the differences between 5C and 6C ULA voltages being due to different peripheral cell resistor values.

| Speaker | MIC | $V_{B2}$ | $V_{C3}$ | A |
|---------|-----|----------|----------|-----|
| 0 | 0 | 0.0302 | 0.0296 | 0.332 |
| 0 | 1 | 0.0302 | 0.478 | 0.744 |
| 1 | 0 | 4.451 | 0.042 | 3.387 |
| 1 | 1 | 4.103 | 2.490 | 3.763 |

Table 20-1: Node voltage SPICE analysis for a 6C001E-7 analogue circuit

| Speaker | MIC | 5C ULA | 6C ULA |
|---------|-----|--------|--------|
| 0 | 0 | 0.391 | 0.342 |
| 0 | 1 | 0.728 | 0.652 |
| 1 | 0 | 3.653 | 3.591 |
| 1 | 1 | 3.790 | 3.753 |

Table 20-2: Voltages produced by ULA speaker and cassette output port

The values predicted in Table 20-1 are very close to the measured values in Table 20-2. The deviation can be attributed to the generic NPN transistor model used for the analysis, as the peripheral cell transistor parameters are unknown.

## Cassette Input

The cassette recorder input is similarly connected to ULA pin 28, through an attenuating resistor and DC blocking capacitor, as shown in Figure 20-1. This input signal alters the bias at junction A, set by the output voltage divider to be at 0.652v when the speaker bit is off and the MIC bit on. This condition is established by the cassette loading routine in the ROM at address 0x05FF, by explicitly setting bit 3 of the ULA output port. The ZX Spectrum ROM disassembly [LOGAN] incorrectly annotates this action as "Signal Mic off", giving rise to the misconception that clearing this bit turns the MIC socket on, and setting the bit turns if off.

The cassette input section of the analogue input/output circuit is presented in Figure 20-3. The voltage divider at the left hand side of the diagram is reproduced from Figure 20-2.

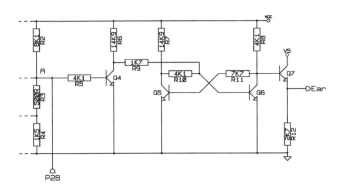

*Figure 20-3: ULA 6C001E-7 cassette input*

The input circuit consists of a single transistor amplifier with a high gain that feeds a bistable multivibrator. The bistable shapes the variable input signal into an absolute high or low logic level as the input crosses a threshold voltage determined by the input amplifier, which reduces sensitivity to noisy signals.

The input amplifier consists of R5, Q4 and R6, and the threshold at which Q4 switches determines the high/low threshold of the input signal. To calculate this, the maximum current that can flow through resistor R6 must be found:

$$I_{Cmax} = \frac{5}{14900} = 0.336 \times 10^{-3} A$$

When Q4 is saturated, $V_{CE}$ = 0, therefore, assuming a transistor gain $\beta$ of 100, the maximum current in the base of Q4 is:

$$I_{Bmax} = \frac{I_C}{\beta} = \frac{0.336 \times 10^{-3}}{100} = 0.336 \times 10^{-5} A$$

The voltage at junction A required to produce a base current of $0.336 \times 10^{-5}$A is the voltage drop across R5 plus the base-emitter voltage drop:

$$V_{IN} = (0.336 \times 10^{-5} \times 4100) + 0.7 = 0.714v$$

As the voltage at junction A gets very close to 0.714v, Q4 rapidly approaches saturation, pulling the voltage at its collector down to 0v. When the voltage at junction A falls away from 0.714v, Q4 stops conducting and allows the collector voltage to be pulled up to Vcc by R6.

The bias at junction A is 0.728v for the 5C ULA and 0.652 for the 6C ULA. The alternating signal fed into the EAR socket, shown in Figure 20-1, causes this bias voltage to rise and fall about its nominal value, crossing the switching threshold as it does so.

When Q4 saturates, the voltage at its collector drops to 0v, pulling the base of Q6 low and turning it off. This allows R8 and R11 to pull the base of Q5 to 0.7v, turning Q5 on and holding the base of Q6 low.

The bistable will remain in this state until the base of Q6 can be pulled high by R6. This occurs when the voltage at junction A drops below 0.714v, shutting off Q4 and allowing R6 to pull the base of Q6 to 0.7v through R9. By turning on, Q6 pulls the base of Q5 low and allows R7 to hold the base of Q6 high. The bistable will remain in this state until junction A once again approaches 0.714v, at which point the bistable will flip states.

The output of the bistable is taken from the collector of Q6 and fed into Q7, an emitter-follower buffer that converts the peripheral cell voltage level (0 to 5v) into a matrix cell compatible voltage between 0 and 0.95v (Vs). This signal is connected to the ULA I/O input port bit 6, designated the EAR input bit, so that the high-low state of the cassette input may be read by the CPU. See the section called *Cassette Recorder Input* in Chapter 19, *Input-Output Devices*.

The cassette recorder signal level will generally exceed 0.7v in amplitude (1.4v peek-to-peek) under successful loading conditions, and because this exceeds the forward voltage drop of the two loudspeaker diodes, the cassette signal will be audible. Though a consequence of having a single multiplexed analogue I/O port, this effect proved beneficial when cueing up programmes to be loaded.

# Chapter 21
# Interrupts

Every task a microprocessor performs may be categorised as transforming an input into an output. Such inputs may come from memory, from storage media or from an external device. Where an input occurs asynchronously, that is, it is not known when it will become available, the processor would need to periodically check for the presence of the input. This wastes time, so some microprocessors provide a mechanism through which to receive notifications that the input device has data to be read. Such notifications are called interrupts, and were discussed previously in the section called *The Z80 Microcomputer* in Chapter 3, *The Standard Microcomputer*.

All microprocessors provide at least one interrupt signal which forces the processor to stop what it was doing and run the Interrupt Service Routine (ISR) associated with the interrupt. The Z80 provides two such interrupt signals: A high priority non-maskable interrupt and a lower priority maskable interrupt.

In addition to informing the processor that an event has occurred, interrupts may be used to provide a time reference or heartbeat against which the processor may mark the passage of time or schedule a regular task. The period of such time signals cannot be very short, as too frequent an interrupt would considerably reduce the performance of the processor in completing its main activity.

## The ZX Spectrum Non-Maskable Interrupts

The Z80's non-maskable interrupt is activated through a dedicated pin labelled /NMI. The processor samples the interrupt signal at the start of the last T-state of the current machine cycle. If it is active, the processor pushes the current program counter onto the stack and jumping to address 0x0066. This interrupt cannot be overridden by software, and takes precedence over the maskable interrupt.

When the ISR has completed processing the interrupt, the return address is popped off the stack and execution is returned to the interrupted program, at the point at which it left off.

The ZX Spectrum does not make use of non-maskable interrupts itself, but provides the Z80 /NMI control signal on its expansion port for peripherals to use. The service routine at address 0x0066 resides in the ZX Spectrum ROM, and was intended to facilitate peripheral use by redirecting NMI execution to a user configurable address. However, a bug in this routine only permits the redirection to address 0x0000, which causes a reset of the computer. Peripheral manufacturers overcome this issue by disabling the ZX Spectrum ROM and enabling their own in its place whenever they activate the /NMI signal.

## The ZX Spectrum Maskable Interrupts

The Z80's maskable interrupt is activated through a dedicated pin labelled /INT and has three modes of operation, selected through dedicated Z80 instructions.

Mode 0 requires that the interrupting devices place an instruction on the data bus for the Z80 to execute. Usually single-byte RST instructions are used, but equally multi-byte instructions are also permitted, such as a jump (JP).

Mode 1 is similar to the non-maskable interrupt in that the Z80 begins execution of an ISR at a fixed address, in this case at 0x0038. This is the default mode used by the ZX Spectrum.

Mode 2 is the most complicated mode and requires that the Z80 I register be set with the upper eight bits of a vector address when the mode is selected. Peripherals complete this vector address by placing its lower eight bits on the data bus whenever they activate the interrupt; therefore allowing multiple devices to have their own service routines and share a common interrupt signal. This vector address does not point to an ISR but instead gives the location of a two byte address within a look-up table. It is the address entries in this table that point to the service routines. This is the mode used by ZX Spectrum software to call their own routines when the ULA interrupt occurs.

The I/O interfaces of the ZX Spectrum, notably the keyboard and cassette interface, are implemented as the simplest form of input and output port, and the ULA does not implement any I/O controller features such as interrupt notification of an I/O event. The ZX Spectrum I/O ports are discussed fully in Chapter 19, *Input-Output Devices*.

To relieve the processor from being put under unnecessary load, and to remove the software burden of having to schedule frequent checks for a keyboard input, the ULA provides the processor with a regular interrupt to be used as a task scheduler. It should be noted that cassette I/O cannot be processed via an interrupt, allowing the processor to simultaneously attend to another task, because the cassette port is sampled by a software routine that requires most of the processor's execution time, and which cannot be interrupted.

When generating an interrupt for the Z80, there is a minimum duration for which the /INT or /NMI signal must be active to guarantee a response from the processor. The Z80 samples both interrupt pins at the rising edge of the last T-state of an instruction, and since the longest instructions take 23 T-states to complete, the interrupt signal must be held low for at least this time.

In choosing a suitable interrupt period, several requirements need to be considered: One, the interrupt must be regular. Two, the period should be short enough to avoid events like key presses from being missed. Three, the frequency of the interrupt cannot be so great as to cause the processor to spend most of its time executing the ISR.

To satisfy these requirements and avoid implementing additional counters to mark out the interrupt period, Altwasser would have looked to see what regular and frequently occurring signal already existed in the ULA. Each stage of the horizontal and vertical counters increasing divide the master 14 MHz clock into longer and longer periods, and the choices would have been clear: Use the period of one scan line, several scan lines, or an entire frame.

The first option of using a single scan line period would result in an interrupt every $64\mu s$, or put another way, after 224 T-states or 56 of the Z80's fastest instructions. The second option of using the period of several scan line, decreases the interrupt period but increases the complexity of the interrupt generator as the number of lines used must be divisible into 312, the number of lines counted by the vertical counter, to make sure that the period was even. The third option of using the period of all 312 scan lines in a single frame gives a period of 19.97ms, which is 69888 T-states or 17472 of the Z80's fastest instructions. Additionally, the ULA is already generating a regular signal at this frequency to provide the television with the necessary vertical synchronisation, and so capitalising on this signal to additionally provide the processor interrupt makes perfect sense.

Because the VSync signal is active for the duration of four scan lines, its additional use as the Z80 interrupt would create a pulse $256\mu s$ long. This would certainly cause repeated detection by the processor since it is very likely that the interrupt signal would still be present when the processor finished executing the service routine, and attempted to resume normal program execution. To avoid this scenario, the ULA creates a shorter interrupt pulse by combining /VSync with V2–0 to reduce its duration to a single 64$\mu$s scan line, and then shortens it further by combining it with the horizontal counter bits C8–6 to create a signal that is active for exactly 32 T-states.

$$\overline{INT} = \overline{VSync} + V2 + V1 + V0 + C8 + C7 + C6$$

It should be noted that the V2 term in the interrupt generation is obsolete, since it also occurs in the VSync logic. Had the vertical sync spanned eight scan lines when the interrupt logic was designed, then the V2 would have been necessary, hinting that the VSync signal timing may have been adjusted during design prototyping. See the section called *Vertical Synchronization* in Chapter 11, *Video Synchronisation* for further information.

$$VSync = \overline{V7 + V6 + V5 + V4 + V3} + V2$$

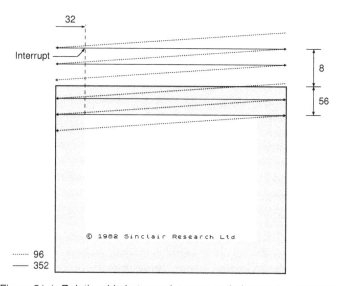

*Figure 21-1: Relationship between interrupt and electron beam position*

As demonstrated by Table 11-1 and Figure 11-4, the horizontal counter is at zero when the electron beam is in line with or at the left hand edge of the pixel display rectangle. Also, the vertical synchronisation pulse begins 64 scan lines before the pixel display rectangle (Table 11-2); therefore the interrupt occurs exactly 64 scan lines before the first pixel of a frame is displayed by the television, which is $64 \times 224$ CPU clock cycles or 14336 T-states. See Figure 21-1.

This timing is exploited by a number of innovative programs that track the exact position of the electron beam through the meticulous counting of T-states, allowing them to change the value of attribute bytes just behind the electron beam, so that the next pixel line of a character row is scanned with a different colour; therefore increasing the vertical colour resolution of that part of the screen by a factor of eight.

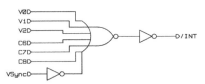

*Figure 21-2: Interrupt signal generation*

The ULA does not place an instruction or vector address on the data bus when it generates the processor interrupt, so interrupt mode 0 cannot generally be used with the ZX Spectrum. Interrupt mode 2, on the other hand, can be used if the vector table is carefully constructed to return the same 16 bit address regardless of which pair of bytes are taken; therefore making the value on the data bus at the time the interrupt was generated irrelevant.

### Detection Reliability

Figure 21-3 shows the relationship between the ULA /INT signal and the CPU clock.

The Z8400A CPU datasheet from Zilog[1] [Z80ZDS] specifies that the minimum */INT fall to clock rise set up time* for the Z80 is 80ns, whereas the timing of the interrupt signal in the ZX Spectrum is such that it occurs approximately 42ns before the next positive clock transition; therefore there is no guarantee that the Z80 will respond to the interrupt at this clock cycle. However, it is known that the processors of most ZX Spectrums *do* manage to recognise the interrupt request at this next cycle. This is borne out first by 14336 being an exact multiple of T-states per line, and second because some ZX Spectrums report 14335 T-states for the same period, which could only occur if the processor completed a cycle before accepting the interrupt request - easily attributed to the out-of-specification set up time.

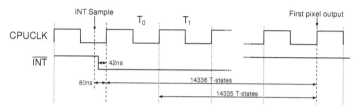

*Figure 21-3: Timing relationship between interrupt and CPU clock*

Why some machines report a "late timing" of 14335 T-states is due to a less tolerant Z80 processor being used, since the ZX Spectrum's interrupt signal is clearly at the limit of detectability by the next clock cycle. These machines have been known to exhibit this timing as soon as they are switched on, or after they have been allowed to warm up. The ULA /INT signal is generated by NOR gating a number of counter signals, and the propagation delay incurred here causes /INT to lag behind the downwards transition of the CPU clock. This lag is increased by the ripple stages within the horizontal counter, delaying the downward transition of the interrupt further so that it is within 42ns of the upward clock transition (measured on a 6C001E-7 ULA). As the ULA warms up, the overall propagation delay experienced by the interrupt increases, closing the gap between it and the rising clock transition and overstepping the detection threshold of a small percentage of processors.

---

1. The Z8400A from Zilog [Z80ZDS], SGS, NEC and the Sharp LH0080 [Z80SDS] all specify a minimum 80ns set up time for the 4MHz processor.

# Chapter 22
# Signal Interfacing

This chapter lists each of the ULA interface pins, giving a brief description of each, whether they are input, output or both, and what electrical interface they implement.

The ULA provides signal level matching between the internal matrix cell voltage levels and the external TTL levels of the other ZX Spectrum ICs, through drivers and buffers implemented within the peripheral cells.

Altwasser made extensive use of standard interface functions from the Ferranti component library, each a pretested component with known operating characteristics. Analogue interfaces such as the YUV video and the cassette/speaker are unique to the ZX Spectrum.

Appendix B, *Component Library* gives each of the interface function schematics.

## ULA Interface Connections

| Pin | Name | Cell | Type | Description |
|-----|------|------|------|-------------|
| 1 | /CAS | 18 | Totem pole output | Column address strobe for the 16K DRAM video memory |
| 2 | /WR | 17 | TTL input | Z80 write enable. Indicates when the processor is performing a write operation |
| 3 | /RD | 16 | TTL input | Z80 read enable. Indicates when the processor is performing a read operation |

| Pin | Name | Cell | Type | Description |
|-----|------|------|------|-------------|
| 4 | /WE | 15 | Totem pole output | Write enable for lower 16K DRAM |
| 5 | A0 | 14 | Tri-state totem pole output; TTL input | Multiplexed lower 16K DRAM address bus; I/O port decoding |
| 6 | A1 | 13 | Tri-state totem pole output | Multiplexed lower 16K DRAM address bus |
| 7 | A2 | 12 | Tri-state totem pole output | Multiplexed lower 16K DRAM address bus |
| 8 | A3 | 11 | Tri-state totem pole output | Multiplexed lower 16K DRAM address bus |
| 9 | A4 | 10 | Tri-state totem pole output | Multiplexed lower 16K DRAM address bus |
| 10 | A5 | 9 | Tri-state totem pole output | Multiplexed lower 16K DRAM address bus |
| 11 | A6 | 8 | Tri-state totem pole output | Multiplexed lower 16K DRAM address bus |
| 12 | /INT | 7 | Open-collector output | Z80 non-maskable interrupt request |
| 13 | Vcc<sub>Logic</sub> | - | | 5v logic supply |
| 14 | Vcc<sub>IO</sub> | 5,6 | | 5v analogue and interface supply |
| 15 | U | 2,3,4 | Analogue | Colour difference signal |
| 16 | V | 45,46,1 | Analogue | Colour difference signal |
| 17 | /Y | 49,50 | Analogue | Video luminance |
| 18 | D0 | 43 | Open-collector output; TTL input | 16K DRAM and Z80 data bus |
| 19 | K0 | 42 | See Keyboard Matrix Inputs | Keyboard matrix half-row column |
| 20 | K1 | 41 | See Keyboard Matrix Inputs | Keyboard matrix half-row column |
| 21 | D1 | 40 | Open-collector output; TTL input | 16K DRAM and Z80 data bus |

230

| Pin | Name | Cell | Type | Description |
|-----|------|------|------|-------------|
| 22 | D2 | 39 | Open-collector output; TTL input | 16K DRAM and Z80 data bus |
| 23 | K2 | 38 | See Keyboard Matrix Inputs | Keyboard matrix half-row column |
| 24 | K3 | 37 | See Keyboard Matrix Inputs | Keyboard matrix half-row column |
| 25 | D3 | 36 | Open-collector output; TTL input | 16K DRAM and Z80 data bus |
| 26 | K4 | 35 | See Keyboard Matrix Inputs | Keyboard matrix half-row column |
| 27 | D4 | 34 | Open-collector output; TTL input | 16K DRAM and Z80 data bus |
| 28 | SOUND | 32,33 | Analogue | Multiplexed cassette I/O and speaker output |
| 29 | D5 | 31,28 | Open-collector output | 16K DRAM and Z80 data bus |
| 30 | D6 | 30 | Open-collector output; TTL input | 16K DRAM and Z80 data bus |
| 31 | D7 | 29 | Open-collector output; TTL input | 16K DRAM and Z80 data bus |
| 32 | /PHICPU | 28 | Inverting open-collector output | Inverted 3.5 MHz clock for the Z80 |
| 33 | /IOREQ | 27 | TTL input | I/O port select |
| 34 | /ROMCS | 26 | Totem pole output | The ROM chip enable |
| 35 | /RAS | 25 | Tri-state totem pole output | Row address strobe for the 16K DRAM video memory |
| 36 | A14 | 24 | TTL input | The Z80 address bus |
| 37 | A15 | 23 | TTL input | The Z80 address bus |
| 38 | /MREQ | 22 | TTL input | The Z80 memory request |

| Pin | Name | Cell | Type | Description |
|-----|------|------|------|-------------|
| 39 | XTAL | 20-21 | Oscillator | Connection to an external 14MHz crystal |
| 40 | GND | 19 | | 0v power connection |

## Interface Categories

### Open-Collector Output

An open-collector output is either at zero volts or in a floating high impedance state. Such outputs are usually connected to an external pull-up resistor, to raise the signal to a high logic level when the output is in its floating state.

The ULA incorporates internal pull-up resistors, but these are generally unnecessary as pull-ups are also provided on the ZX Spectrum PCB.

### Totem Pole Output

A totem pole output contains two output transistors, one that pulls the signal up to a logic 1, the other that pulls the signal down to a logic 0. They are typically used as output stages of TTL integrated circuits, and sometimes incorporate an output enable that switches off both output transistors when not active, disconnecting the output from the external circuit.

### TTL Input

A TTL input is one that is compatible with the TTL logic 1 and 0 voltage levels. The ULA implements this through a voltage divider and emitter-follower that has its collector connected to Vs. This converts the TTL level input signal into an internal CML logic compatible signal.

## ULA Interface Summary

### Z80 Control Signal Input

The Z80 control inputs are accepted by TTL level compatible emitter-follower buffers, where they are converted into CML matrix cell compatible logic levels.

## Z80 Address Bus A15-14

The address bus inputs A15-14 are accepted by TTL level compatible emitter-follower buffers, where they and converted into CML matrix cell compatible logic levels.

## Address Bus A6-0

The ULA generates six multiplexed DRAM address bus signals, through which it selects the video memory address it wants to read. These outputs are tri-state and are only enabled during a video update memory fetch, when internal signal /AE is active. Details of the video address generation and bus enable can be found in the section called *Generating The Address* in Chapter 15, *Video Addressing*.

A0 is a special case, being both an output and an input. When the ULA address bus is disabled, the Z80 A0 signal is passed to pin 5 through the external address multiplexer and bus isolating resistor. The ULA buffers this signal, converting it to a CML compatible logic level that is used by the I/O port decoding described in the section called *Decoding the I/O Port* in Chapter 19, *Input-Output Devices*.

## Data Bus D7-0

As the ULA both reads and writes to the data bus, this interface consists of non-inverting collector-follower outputs with weak pull-up resistors. Coupled to these are emitter-follower input buffers that convert the external TTL levels into CML compatible logic levels. Data bus signal D5 is an exception, as it is input only.

Weak pull-up resistors are used because the data bus is shared with other ICs such as 16K and 32K DRAM, ROM and external peripherals. The ZX Spectrum provides additional pull-up resistors on the PCB.

## Z80 Interrupt Signal

The ULA generated interrupt signal for the Z80 is converted to a TTL compatible level by a non-inverting collector-follower. The output has a strong pull-up resistor to provide a fast rise/fall time.

## PHICPU 3.5 MHz Z80 Clock

The ULA generated Z80 clock is converted to an inverted TTL compatible output by a collector-follower. The output incorporates a strong pull-up resistor to provide a fast rise/fall time.

## Keyboard Matrix Inputs K4-0

The keyboard matrix inputs are passed into the ULA through an emitter-follower buffer with a weak pull-up resistor, converting the external voltage levels into CML compatible logic levels.

Keyboard pin K0 is an exception because it also provides an open-collector output. This output may pull K0 down to 0v as part of ULA die testing during manufacture. See the section called *V8 Output Via Keyboard Input K0* in Chapter 23, *Hidden Features and Errors* for further details.

When a key is pressed, the Z80 address bus pulls the keyboard input signal down towards 0v by sinking current. Resistance due to the keyboard contacts and the connecting cable reduces the voltage swing at the ULA input pins. The input buffers take this into account and are biased differently to the TTL inputs, accepting a smaller voltage range.

## ROM Chip Enable

The ROM chip enable is a totem pole output that goes low when the CPU requests access to the 16K ROM.

## DRAM Column Address Strobe

The DRAM CAS is a totem pole output that goes low as part of the video generator or Z80 access to the lower 16K DRAM.

## DRAM Row Address Strobe

The DRAM RAS is a tri-state totem pole output, enabled when either the video generator or Z80 requires access to the lower 16K DRAM. There is an implementation error with this signal which results in it being enabled for most of the time. For further details, see the section called *Disabled 16K DRAM Refresh* in Chapter 23, *Hidden Features and Errors*.

# Chapter 23
# Hidden Features and Errors

## Test Modes

A significant cost in the production of integrated circuits is the time taken to complete wafer testing. Each die on a wafer is tested against a number of customer supplied test vectors, using an automated wafer probe that connects to the die pin connections. The die that fail are spot marked with ink so that they may be identified and discarded before packaging. If test vectors take a long time to complete, manufacturing costs will rise considerably.

It is therefore imperative that customers ensure the test pattern will execute in the shortest time possible. For some designs, this may involve building specific test features into the core functionality of the device.

The ZX Spectrum ULA contains four such test features that allow die testing to exercise its complete clock and signal cycle in a compressed time period.

The issue with testing the ZX Spectrum ULA is that it is driven by a single clock signal, the frequency of which it repeatedly halves to generate the scan line and video frame timing.

Under normal operating conditions, the ZX Spectrum ULA is clocked at 14MHz and completes a full video cycle in 20ms. This is an exceptionally long time when die testing, and even by clocking the ULA at the maximum permitted frequency of 20MHz, completion time only improves to 14ms.

Altwasser realised that there were key clock signals in his design that would allow rapid test pattern execution if they could be driven at a much higher frequency than usual. To allow this, he identified ULA pin signal combinations that would normally be invalid, and engineered the detection of these invalid conditions so that they overrode the required internal clocks.

Essentially, it was not necessary to test an entire frame. As long as the key operations performed during a scan line and frame were checked, the correct operation of the ULA video handling would be proven.

Checking events within a frame, such as the vertical sync and CPU interrupt, was sped up considerably by allowing the upper stage of the master counter to be clocked directly. This allowed all 312 states of the vertical counter to be stepped through in 0.1092ms, and is the first of the test mode clocks.

Stepping through the events of a scan line was achieved by clocking the master oscillator input of pin 39 at 20MHz; the complete sequence taking 0.0448ms. The test vectors would verify the timing of signals such as /RAS, /CAS, /HSync and pixel video generation, having pre-loaded the ULA data bus inputs with a suitable test pattern.

Several aspects of the pixel video generation needed to be tested independently, such as the PAL odd/even line inversion, the flash counter and flash mode colour inversion. This required clocking the flash counter much faster than usual, the ability to measure the RGB values after the PAL line inversion, and some indication as to the current scan line number.

The flash counter clock is the second test mode clock, and is discussed below along with the additional test monitoring signals.

To enable the test modes, the ULA decodes CPU signals /MREQ and /IOREQ, looking for the invalid condition of both being low together, and addresses each clock through the /RD and /WR signals.

**Upper Counter Stage Test Clock**

The first test clock /TCLKA is defined as:

$$TCLKA = \overline{\overline{IOREQ} + \overline{MREQ} + \overline{RD} + WR}$$

In the section called *The Master Counter* in Chapter 10, *The Internal Clocks*, it was discussed that the master counter segments C8–6 and the entire vertical line counter are clocked by CLKHC6, the result of gating /C5 with /TCLKA:

$$CLKHC6 = \overline{\overline{C5} + \overline{TCLKA}}$$

Under normal operating conditions /TCLKA will be low, and clock signal CLKHC6 reflects the state of C5. However, when the test condition is imposed by /MREQ, /IOREQ and /RD being low, /TCLKA will go high; therefore CLKHC6 reflects the state of TCLKA, when /C5 is low.

To perform these accelerated tests, the master clock is removed from the oscil-lator input pin 39, and /MREQ and /IOREQ are held low to activate the clock override. /RD is then clocked at up to 20MHz, directly driving CLKHC6 at this frequency. As the override will only succeed when master counter /C5 is low, the tester would guarantee that /C5 was low before it proceeded with the test.

It is probable that the scan line tests for signals such as /RAS, /CAS and hori-zontal sync would be performed first, and in an order that results in /C5 being left low. See Table 11-1.

CLKHC6 drives the upper three segments of the master counter, which cycle through seven states before clocking the vertical counter. It is therefore possi-ble to clock the vertical counter at 20MHz / 7 = 2.86MHz instead of the usual 15.625Hz, sequencing a single video frame in 0.1092ms instead of 19.968ms.

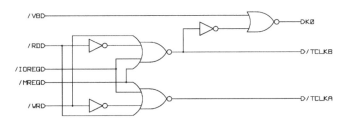

*Figure 23-1: TCLKA and TCLKB generation, K0 output*

### Flash Counter Test Clock

The second test clock /TCLKB is defined as:

$$TCLKB = \overline{IOREQ + MREQ + RD + WR}$$

This clock is similar to the /TCLKA discussed above, and is enabled under the same invalid conditions of /IOREQ and /MREQ being low together, with /WR being the clock input instead of /RD.

This test clock overrides /V8 which clocks the flash counter to generate the 1.565 Hz flash signal (see Figure 14-3), allowing the flash counter to be clocked at 20MHz, producing a 625KHz flash signal. The test process would remove the master clock from the oscillator input pin 39, when clocking /TCLKB input /WR.

Because /V8 must be low for /TCLKB to control the flash counter, the ULA provides a mechanism for the wafer probe to monitor V8, see the section called *V8 Output Via Keyboard Input K0*.

Driving the flash counter directly allows rapid testing of the flash counter, flash XNOR and colour output multiplexer by monitoring the /Black* output, see the section called *Black Output*.

Before carrying out these tests, the flash mode must be enabled by presenting an appropriate value to the ULA data bus during the video byte load test (see the section called *The Flash Mode* in Chapter 12, *Generating The Display*).

### V8 Output Via Keyboard Input K0

While /TCLKB is enabled and high, V8 is output through keyboard input connector K0.

$K0_{out}$ is given by:

$$\overline{K0_{out}} = \overline{V8} + TCLKB$$

The KO output is an open collector, so normally it has no effect on the incoming keyboard signal. However, when $K0_{out}$ goes high, the output of K0 goes low as it is clamped to ground.

This output is used during die tests to identify when /V8 is low, so that /TCLKB may take control of the flash counter, and also to indicate when the first vertical counter has reached the first pixel display line.

### Black Output

/Black* is routed to the unused peripheral cell 48, where it is output to the package pin bond pad via transistor T6, forming an open collector output.

This bond pad provides the NOR of the colour output multiplexer RGB values, after they have been processed by the PAL odd/even line inverter. By sampling the value at this connection, the die test pattern can test whether or not black is being produced when expected, whether the flash counter and XNOR gate are functioning correctly and whether the PAL odd line inversion is occurring appropriately.

Altwasser's patent *Display for a computer* [ALTWASSERDC] shows the Black* signal being inverted and fed off the bottom of Fig. 5D.

## Implementation and Design Errors

### Disabled 16K DRAM Refresh

The lower 16K DRAM refresh mechanism of the issue 1 ZX Spectrum intended to rely on the ULA video access during the active part of the display, and the Z80 refresh mechanism at all other times.

Thus the /RAS output of the ULA is supposed to go into a high impedance state when a video update or Z80 16K DRAM access is not being performed. This hands control of the DRAM /RAS input over to the Z80 /RFSH signal, which is also connected to the DRAM /RAS via a 330R resistor, as in Figure 23-2.

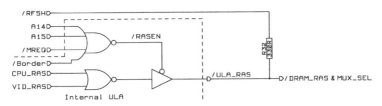

*Figure 23-2: ZX Spectrum /RAS output and enable*

As the implementation stands, the /RAS output of the ULA is a tri-state signal that is enabled when:

$$\overline{RASEN} = \overline{MREQ} + A14 + A15 + \overline{Border}$$

This shows that the /RAS output is deactivated and goes into a high impedance state only when the Z80 is addressing the ROM during periods of border update. At all other times /RAS will be enabled, overriding the Z80 /RFSH signal. In other words, the Z80 can only refresh the 16K DRAM while it is accessing the ROM *and* the video controller is not updating the display.

The ZX Spectrum suffers no ill effects from the disabled refresh as the video update appears to provide enough of a row refresh for the 4116 DRAM chips to maintain integrity. As the ULA does not access the video DRAM for 120 consecutive display lines, the longest period between DRAM memory row reads is 7.68ms, which exceeds the maximum period of 2ms specified by the data sheet [DS4116].

For the ULA /RAS output to go high impedance whenever the video controller and CPU are not accessing the 16K DRAM, /RASEN should have been:

$$\overline{RASEN} = \overline{(MREQ \cdot A14 \cdot \overline{A15}) + Border}$$

$$\equiv \overline{(\overline{MREQ} + \overline{A14} + A15) + \overline{Border}}$$

$$\equiv \overline{RAM_{16} + \overline{Border}}$$

This logic was not fixed in subsequent revisions of the ULA, and the /RFSH connection to /RAS via R32 was removed from the issue 2 ZX Spectrum and above.

## Dark Flash Edges

Display artefacts that all ZX Spectrum owners will be familiar with are the vertical lines that appear at the left and right edges of flashing character cells. An example of this is shown in Figure 23-3.

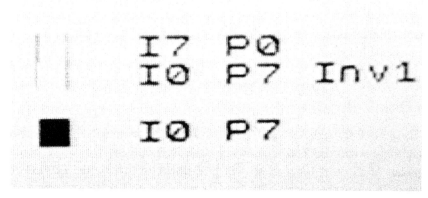

*Figure 23-3: The dark edges of flash cells*

Consider the first character cell shown in the example. It has a vertical line at its left and a thinner vertical line at its right. This empty character cell has its flash mode enabled and has a black background (paper) and a white foreground (ink) attribute. Because this photograph was taken while the character cell was inverted by the flash clock, the cell is shown with a white background.

The character cells to the left and right of the flash cell have white background and black foreground attributes.

These vertical lines are due to the flash mode enable signal being delayed between the colour output latch and the DataSelect XNOR gate. This is demonstrated by the following sequence of events, which assume that the FlashClock signal is low:

1. As the electron beam moves from the non-flash character cell on the left, into the flash cell, the white foreground and black background colours of the flash cell are presented to the colour output multiplexer.

   The flash mode enable signal, FL, is applied to the XNOR gate shown in Figure 23-4, but is delayed by an inverter and NOR gate.

   Because of the delay, the expected inversion of /DataSelect does not occur at the XNOR gate at this time, causing the background colour to be selected by the output multiplexer, and the television to being displaying black.

2. After the delay of the flash enable signal has past, the XNOR gate inverts /DataSelct and causes the output multiplexer to correctly select the white foreground colour. The television will begin to display white.

   This ends the moment of black output, leaving a vertical line to the left of the character cell.

3. As the electron beam enters the non-flash character cell to the right of the flash cell, the colours presented to the output multiplexer are set back to a foreground of black and a background of white.

   The flash mode enable signal is disabled, but is again delayed before it reaches the XNOR gate.

   /DataSelect continues to be inverted and the foreground colour of black begins to be displayed by the television.

4. Shortly after this, the disabled flash mode signal reaches the XNOR gate, reverting /DataSelect and causing the output multiplexer to select the background colour, white.

   This ends the moment of black output, leaving a vertical line at the right hand edge of the character cell.

The example above works by swapping the foreground and background colours as the electron beam enters and leaves the flash cell. The DataSelect signal does not change state because all pixels are reset.

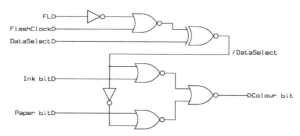

*Figure 23-4: Flash XNOR and foreground/background colour bit multiplexer*

The vertical lines can also be generated by keeping the foreground and background colours consistent, and filling the flash cell completely with set pixels. However this will produce a thinner left hand vertical line than the previous example. The second example in Figure 23-3 shows a flash cell with inverted pixels, so that they are all set:

1. As the electron beam moves from the non-flash character cell on the left, into the flash cell, the set pixels cause DataSelect to go high.

   The flash mode enable signal, FL, is applied to the XNOR gate shown in Figure 23-4, but is delayed by an inverter and NOR gate.

   Because of the delay, the expected inversion of /DataSelect does not occur at the XNOR gate at this time, causing the foreground colour to be selected by the output multiplexer, and the television to being displaying black.

2. After the delay of the flash enable signal has past, the XNOR gate inverts /DataSelct and causes the output multiplexer to correctly select the white background colour. The television will begin to display white.

   This ends the moment of black output, leaving a vertical line to the left of the character cell.

3. As the electron beam enters the non-flash character cell to the right of the flash cell, the reset pixels cause DataSelect to go low.

   The flash mode enable signal is disabled, but is again delayed before it reaches the XNOR gate.

/DataSelect continues to be inverted and the background colour of black begins to be displayed by the television.

4. Shortly after this, the disabled flash mode signal reaches the XNOR gate, reverting /DataSelect and causing the output multiplexer to select the foreground colour, white.

This ends the moment of black output, leaving a vertical line at the right hand edge of the character cell.

The flash cell in the third example of Figure 23-3 is identical to the first example, except that the flash cell has a foreground of black and a background of white. Vertical edges have been produced, but as they are white they blend in to the overall screen background.

Examining the lines generated by examples one and two, it is clear that the line at the left hand edge of the first example is thicker than the other three lines. This appears to be due to the slightly asymmetrical propagation delay of the foreground/background colour bit multiplexer. See Figure 23-4.

When the ink (foreground) bit is clear, and the paper (background) bit is set, the propagation delay of the multiplexer is the sum of the ink and output NOR gate propagation delays. When the ink bit is set and the paper bit is reset, the propagation delay of the multiplexer is the sum of the channel select inverter, paper and output NOR gate propagation delays.

Where the first example demonstrates a thicker line at the left edge of the flash cell, the multiplexer is taking a little longer to select the foreground colour of white because it is slower when switching between a set foreground bit and a reset background bit.

**Variable Pixel Widths**

The photograph in Figure 23-5 show an alternate pixel pattern that is darker in the second and fourth columns. The first eight character cells from left to right are displaying a 1 0 1 0 1 0 1 0 pixel pattern, with a black foreground and white background. This pattern is repeated down the screen, creating the lighter shade column on the left.

To the right of this column, character positions eight to fifteen are filled with an inverted pixel pattern of 0 1 0 1 0 1 0 1, but this time with a white foreground and black background. As the colours have also been inverted, the television displays the same 1 0 1 0 1 0 1 0 pattern as the first column, but appears darker than the first. The two columns are repeated once more, giving four wide columns in total.

Columns two and four have a black background and a white foreground. As described in the section called *Dark Flash Edges*, this combination of foreground and background colour increases the propagation delay of the foreground/background colour bit multiplexer, widening the area of colour being switched from. However both the foreground and background colours would be delayed equally, and there would be no visible brightness difference.

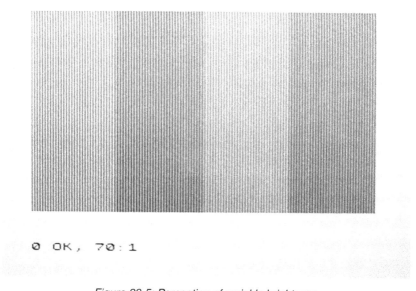

*Figure 23-5: Perception of variable brightness*

It is probable that the large fan-out of the /DataSelect signal is affecting its switching speed, such that it is able to pull a gate input high in less time that it is able to pull it low. This would make the reset background pixels very slightly wider than the set foreground pixels, the visual effect of which would be amplified by alternating between set and reset pixels. In the example shown in Figure 23-5, the asymmetric pixel width would lighten columns one and three, and darken columns two and four, the contrast between the two appearing much greater.

## All I/O Ports Contended

This has already been discussed in the section called *The issue 2 5C112 ULA* in Chapter 18, *CPU Clock and Contention*, but will be summarised here.

The CPU contention handler does not consider A0 along with /IORQ when deciding whether the Z80 I/O activity will cause contention for the ULA data bus; therefore *all* I/O activity is considered equally, resulting in the processor being stopped when *any* I/O port is accessed during a display update.

Sinclair engineers solved this problem without needing to produce a replacement ULA by OR gating A0 and /IORQ on the PCB and feeding the result into the ULA as /IOREQ. This OR gate was created by the resistor that was already present in the /IORQ signal and an additional transistor soldered across the Z80 between Vcc, A0 and /IOREQ (at the ULA side of the resistor). See Figure 18-13.

## Colour Sub-Carrier Frequency Lock

During the design of the ZX Spectrum, Sinclair engineers were unaware that PAL and NTSC TV standards required the colour sub-carrier frequency to be locked to the line and frame frequencies, and consequently designed the colour encoder to be clocked from a different frequency source than the ULA, which is responsible for the synchronisation signals. The cross-talk effect of these unlocked frequencies is to produce a thin vertical rolling effect between the vertical edges of strongly contrasting colours.

## Phantom Keys

This feature is not really a design flaw of the ULA, but arises due to the way the keyboard is multiplexed through the ULA's single I/O port.

When software checks for a keypress in the half-row Q to T, it reads from I/O port 0xFB7F. See Table 19-1. This pulls address line A10 low to select the desired keyboard half-row, and reads the 5-bit keypress result from the ULA I/O port.

If the user pressed Q, P and O simultaneously, the read of keyboard half-row Q to T would incorrectly indicate that Q and W were being pressed.

What is happening here is that by pressing the Q key, A10 is connected to K0, which pulls K0 low. By pressing P at the same time, A13 is also connected to K0, which causes A13 to be pulled low. Now, by pressing another key connected to A13, in this example O, its keyboard input, K1, is mistakenly pulled low by A13; therefore a read of half-row Q to T sees K0 and K1 going low, which is interpreted as keys Q and W being pressed.

## The Snow Effect

The "Snow Effect" occurs because the ULA does not consider a refresh address between 0x4000 and 0x7F7F as a contention condition, and does not halt the instruction being executed while it completes the video byte fetch. See Chapter 18, *CPU Clock and Contention*. While this is technically correct, the CPU /RAS generator recognises this condition as being a genuine access to the lower 16K DRAM, and therefore generates a CPU /RAS signal. This interferes with the /RAS signal being produced by the video generator, and corrupts the video byte fetch that is taking place. See the section called *CPU RAS Generation* in Chapter 17, *CPU Memory Access* for further information.

Detection of a conflicting CPU address only occurs during the first T-state of an instruction. The second T-state is identified by the rising clock transition after /MREQ has gone low, and the last T-state by the rising clock transition after /MREQ has gone high. See Figure 23-6. During this period contention checking is disabled to allow the instruction to complete, and is controlled by signal MREQT23.

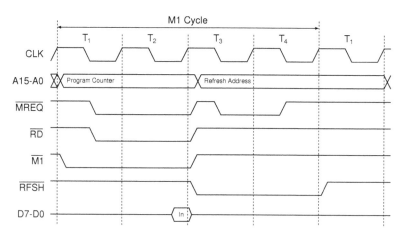

*Figure 23-6: Z80 instruction fetch showing refresh cycle*

If an instruction is fetched from the lower 16K DRAM, the address will be detected by the ULA during the first half of $T_1$, and the instruction will be held until the video fetch is complete. However, because an instruction fetch is four T-states in length, the timing of the clock wait signal, CLKWAIT, is such that $T_4$ can overlap the start of the video fetch, where /VidRAS goes low.

Furthermore, by ignoring the address bus between the start of T-state $T_2$ and the end of $T_4$, the ULA will not consider the refresh address placed on the bus during $T_3$ and $T_4$.

During a refresh cycle, the Z80 places the lower seven bits of the R register on address lines 6-0. In addition, it places the contents of the I register on the upper eight address lines, an action of the Z80 that is not officially documented by Zilog. Therefore, if the I register is set to a value between 0x40 and 0x7F, the CPU will generate a refresh address between 0x4000 and 0x7F7F.

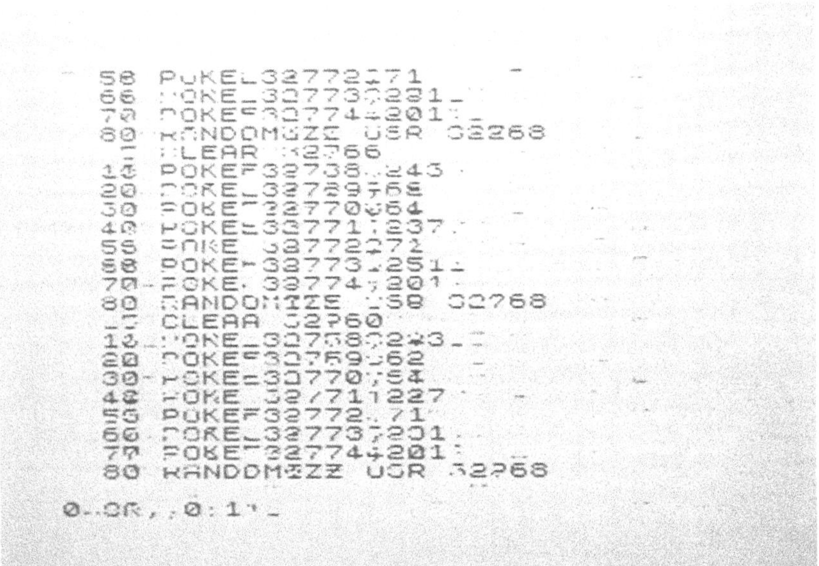

*Figure 23-7: Screen break-up of the "Snow Effect"*

The consequence of not considering the refresh address during contention checking allows the instruction fetch refresh cycle to overlap the video byte fetch. If the refresh address is between 0x4000 and 0x7F7F, the ULA will generate a CPU /RAS signal and disrupt the timing of the video /RAS. This can cause /RAS to stay low across both page mode reads, or cause /RAS to go low before the video DRAM row address has been placed on the ULA address bus. Both of these situations lead to video bytes being fetched from the wrong row address, resulting in a break-up of the screen. See Figure 23-7.

Had the Z80 /RFSH signal been available to the ULA, the generation of the CPU /RAS signal during a refresh cycle could have been suppressed. Alternatively, /RFSH could have been used on the ZX Spectrum PCB to pull /MREQ high at the ULA pin, preventing the CPU /RAS from being produced.

# Chapter 24
# ULA Versions

## Issue 1 ULA 5C102E

The first ZX Spectrum ULAs, the 5C102E, were produced by Ferranti from March/April 1982. The earliest recorded date code is 8214, giving the date of manufacture as the week beginning 4 April 1982. Many of these early machines were sent out as review models, with the operating system contained in an EPROM as the ROMs had not yet been produced.

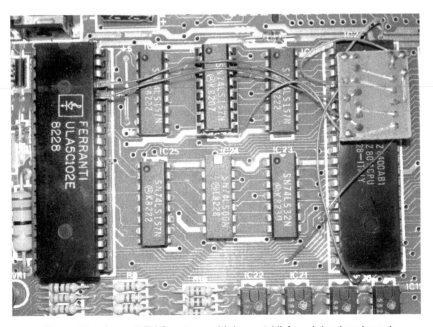

*Figure 24-1: Issue 2 ZX Spectrum with issue 1 ULA and dead cockroach*

The contention handling of this ULA contained a logic error that prevented I/O contention from being handled properly. This only partially halted the execution of I/O instructions while the video update was being performed, and was first noticed by the erratic keyboard response of the first machine code games. See the section called *The issue 1 5C102 ULA* in Chapter 18, *CPU Clock and Contention* for full details.

The engineers at Sinclair corrected the logic error by modifying the ZX Spectrum PCB circuit with a small upside-down IC, termed the "dead cockroach". See Figure 18-9.

Some early issue 2 ZX Spectrum models were sold containing the issue 1 5C102E ULA and dead cockroach modification, mounted on its own tiny PCB. The photograph in Figure 24-1 shows such a machine, serial number 001-037728, containing an issue 1 ULA manufactured between the 21 and 27 June 1982.

The erratic I/O contention handling masked a second error that caused the ULA to interrupt *all* Z80 I/O operations, even if they were not intended for the ULA. This error remained undiscovered until after the issue 2 ULA went into production.

## Issue 2 ULA 5C112E

The issue 2 ULA, 5C112E, was first manufactured circa August 1982. It corrected the earlier I/O contention error of the ULA 5C102E by including the "dead cockroach" modification internally, however the second I/O handling error present in the 5C102E ULA remained unnoticed and was therefore not fixed.

The contention handler does not consider A0 along with the Z80 /IORQ signal during I/O activity, and therefore treats all I/O requests equally, interrupting requests for I/O ports such as the ZX Printer.

The Sinclair engineers corrected this problem by OR gating A0 and /IORQ together before passing them to the ULA /IOREQ pin. This was neatly achieved by soldering a transistor "spider" across the top of the Z80. See photograph in Figure 18-14.

It was also discovered that the intended Z80 memory refresh of the lower 16K DRAM did not work as intended with the issue 1 ZX Spectrum, and caused no ill effects. Therefore the /RFSH signal was disconnected from the lower 16K /RAS signal on the issue 2 ZX Spectrum PCB, and was not corrected within the 5C112E ULA. Presumably this fault was discovered after the 5C112E had been produced and was cleaned up during the issue 2 ZX Spectrum PCB redesign.

## Issue 3 ULA 6C001E-6

The issue 3 ULA, 6C001E-6, was first manufactured circa May 1983. It was based on a new ULA series from Ferranti, and may have been produced specifically for Sinclair Research.

This new series used Ferranti's improved ULA manufacturing process, resulting in a lower power device. Resistor changes on the PCB were therefore necessary when using the 6C001 ULA, resulting in a PCB design change for the issue 3 ZX Spectrum and above.

The ULA cleaned up the earlier I/O contention fixes that had been applied to the issue 2 5C112E ULA and altered the timing of the colour burst component of the video signal, improving compatibility with Hitachi and Grundig televisions. This had the side effect of shifting the television picture slightly further to the left of the screen. See the section called *Burst Generation* in Chapter 16, *Analogue Video* for full details.

The issue 2 "spider" modification was still required, and became an integral component of the ZX Spectrum PCB.

## Issue 4 ULA 6C001E-7

The issue 4 ZX Spectrum contains a modification that improves the reliability of the CPU's access to the lower 16K DRAM.

The multiplexer select signal in earlier issues of the computer is taken directly from the DRAM /RAS signal, switching the multiplexers over to the column address when it goes low. This removes the row address soon after /RAS goes low, potentially violating $t_{RAH}$. See Table 13-1 and Figure 13-1.

In the issue 4 ZX Spectrum and above, the address bus multiplexer select signal is delayed by two NAND gates on the PCB, connected as inverters. This increases the length of time between /RAS going low and the removal of the row address, improving row address reliability. The unfortunate side effect of delaying the column address is that the address bus may not stabilise before /CAS goes low, making the column address read unreliable. The 6C001E-7 ULA was produced to solve this problem by generating a delayed /CAS signal.

The ULA introduces delay into the /CAS signal with two inverters, shown between /comCAS and /CAS in Figure 17-7. This was the most convenient location to insert the delay, and required minimal interconnect re-routing. This affects both the CPU CAS and the video CAS, but was only necessary for the CPU CAS.

For example, the CPU generated /CAS signal goes low 94ns after /RD for the 6C001E-7 ULA, and after 61ns for the 6C001E-6. Similarly, the video generated /CAS signal goes low approximately 78ns after /RAS for the 6C001E-7 ULA, and 50ns after /RAS for the 6C001E-6 ULA. Therefore, CAS timings are approximately 30ns later for the 6C001E-7 ULA.

Because the PCB NAND gates alter the timing of the multiplexer select signal, it is important that the /CAS timing is also adjusted by using a 6C001E-7 ULA is these machines. Using an earlier ULA will cause the computer to behave unreliably. This is reflected by the ZX Spectrum Service Manual Supplement No 1, which state that "Two spare gates from IC24 have been used to improve ULA timing. It is important that ULA 6C001-7 or later issues (the issue number is designated by the number after the hyphen) are used in this circuit. The gates are fitted in series with the RASL output from the ULA and the input to the display RAM multiplexers."

The 6C001E-7 ULA can be used successfully in earlier issues of the ZX Spectrum.

## The NTSC ULA 6C011E

The NTSC ULA, identified as 6C011E, was produced by Sinclair presumably to export the ZX Spectrum to the United States. However, the design of the machine did not comply with the FCC regulations for computers, and its sale was prohibited. The earliest known date code for an NTSC ULA is 8444, giving the date of manufacture as the week beginning 29 October 1984. The computer containing this ULA was sold in Chile, South America, and it is not known how many were manufactured.

The NTSC ULA is clocked by a 14.11 MHz crystal, slightly faster than a PAL ULA, producing an NTSC compliant $63.5\mu s$ scan line. At 264 lines per frame, the ULA generates a frame rate of 59.65 MHz. The Z80 CPU runs slightly faster as a result, being clocked by the ULA at 3.5275 MHz. See the section called *The NTSC Display* in Chapter 9, *The Video Display* and the section called *The NTSC Line Counter* in Chapter 10, *The Internal Clocks*, for further details.

The vertical synchronisation pulse occurs earlier than for a PAL ULA, 216 scan lines after the first pixel display row, shortening the bottom border to 24 lines. This timing adjustment helps maintain a vertically centred display. See the section called *NTSC Vertical Synchronisation* in Chapter 11, *Video Synchronisation*.

Colour generation is also affected, as the ULA does not invert the chrominance V signal on alternate scan lines, and only inserts a colour burst into the U signal. Additionally, the LM1889 chrominance modulator on the ZX Spectrum PCB is clocked by a 3.579545 MHz crystal, providing the necessary NTSC colour sub-carrier frequency. See the section called *NTSC Chrominance Modulation* in Chapter 16, *Analogue Video*.

## The ZX Spectrum 128 ULA

The ZX Spectrum 128 was produced by Sinclair Research in partnership with its Spanish distributor, Investrónica. The design of the 128 ULA, designated 7K010E-5, is almost identical to the 5C and 6C ULAs used in the ZX Spectrum 16/48K, and contains very few design changes or fixes. For example, the "spider" modification is still required to prevent the ULA contending all I/O operations. The main modifications to the ULA were to improve the quality of the video signal generated, and to provide additional clocks for new circuit components.

In the section called *Colour Sub-Carrier Frequency Lock* in Chapter 23, *Hidden Features and Errors*, the vertical rolling effect that occurs between strongly contrasting colours is described. This is caused by interaction between the two separate oscillators used for the video generation. The 7K010E ULA avoids this by using a single crystal oscillator from which it derives the master counter clock, and the colour sub-carrier signal.

The 7K010E ULA does not produce luminance and chrominance YUV signals, but instead outputs the internally generated RGB, Bright, VSync and composite synchronisation signals through six ULA package pins. These video components are modulated into a composite video signal by a TEA2000 colour modulator on the PCB, in addition to making them available at an RGB monitor connector at the rear of the computer. The TEA2000 modulator requires an external clock at twice the PAL sub-carrier frequency of 4.43361875 MHz.

The ULA is driven by a 17.7345 MHz master clock, the frequency of which was carefully chosen to allow both the colour sub-carrier and pixel clock to be generated by the ULA. By dividing the master clock by 2, the ULA generates the 8.86725 MHz colour clock for the TEA2000, and by dividing by 2.5, it generates the 7.0938 MHz pixel and master counter clock. This pixel clock is slightly faster than used by the 16/48K ULA and, without modifying the master counter, would result in a shorter scan line period of $63.15\mu s$.

The horizontal and vertical line timings of the 7K010E ULA have therefore been adjusted to take account of this higher clock frequency. For instance, the number of master counter states that produce a 15625 Hz line frequency is given by:

$$\frac{7.0938MHz}{15625Hz} \approx 454$$

However, in the 7K010E ULA, the master counter responsible for tracking the horizontal position of the television electron beam is reset on reaching 456, and not 454 as predicted above. The upper six bits of the counter, C8-3, must therefore be constructed from the larger T-Type flip-flops which have reset, carry and enable, so that the counter stages are synchronous with a common reset signal and clocked by C2. The reset signal is activated when the counter reads 455 (111 000 111), and the reset occurs at the next clock transition. The reset is therefore generated from the NOR of /C8, /C7 and /C6 and is applied until C2 advances from 1 to 0, clocking the synchronous stages as it does so and resets them to 000000.

Establishing the reset at 456 meant that the C2-0 did not need to be considered in the generation of the reset signal, and could remain as D-Type flip-flops, saving space. The consequence being that the counter generates a horizontal line period of $64.28\mu s$.

With this slightly over-specification line period, the frame rate would become:

$$312 \times 64.28 \times 10^{-6}s = 20.05 \times 10^{-3}s = 49.86Hz$$

This is below the specification 50Hz, and so the engineers of the 7K010E ULA reduced the vertical line count to 311 to compensate:

$$311 \times 64.28 \times 10^{-6}s = 19.99 \times 10^{-3}s = 50.02Hz$$

The increase in master counter frequency increases the CPU clock to 3.5469 MHz, a gain of 46.9 KHz.

In addition to generating the pixel, colour sub-carrier and CPU clocks, the ULA generates the 1.77345 MHz clock for the AY-3-8912A sound chip by dividing the CPU clock by two and passing this out of a dedicated package pin. All other aspects of the ZX Spectrum 128, such as the 128K bank-paged memory and the control of the AY-3-8912 sound device are handled by a custom PAL chip, and the custom ZX8401 multiplexer that also replaced the two 74LS157 multiplexers in the ZX Spectrum 48K issue 5 and above.

The cassette interface described in Chapter 20, *Cassette Storage and Sound* is also implemented across two peripheral cells in the 7K010E ULA. However, instead of both input and output interfaces being connected to a single ULA pin, the MIC and Speaker output section (Figure 20-2) is connected to package pin 35, and the EAR input section (Figure 20-3) is connected to package pin 34. On the ZX Spectrum 128 PCB, these two pins are connected together, reproducing the behaviour of the original ZX Spectrum 16/48K.

The 7K010E-5 ULA identifier may be broken down as follows. 7 indicates that it is a 7000 series array, K gives the array type and speed, 010 the device customisation identifier, E the package type (plastic DIL in this case) and 5 the revision number. Like the 6000 series array, the 7000 series does not appear in the Ferranti archives held at the Museum Of Science and Industry in Manchester. It is likely that the 7000 series is a larger version of the 6000 series, with the same matrix and peripheral cell structure, and the same component values; just as the 5000 series array is a larger version of the 2000 series. This would have been desirable to reduce the amount of redesign and re-layout required and to reduce cost. The 7000 series array was the largest used by Sinclair, and has 48 package pins.

# Appendix A
# The ULA Die Plot

The die plot of the ULA 6C001E-7 silicon is presented in Figure A-1 and Figure A-2, and shows the location and size of each functional design component within each ULA quarter.

The first quarter contains the analogue video signal and colour generation, along with the flash counter and the upper master and lower vertical counter stages.

The second quarter contains the video address generation, master and vertical counter stages and I/O port address decoding.

The third quarter contains the display and attribute byte data latches, attribute output latch, colour output multiplexer and partial pixel shift register. Also present are the IO port output latches, keyboard input buffers and cassette recorder EAR input buffer.

The fourth quarter contains the master clock, lower master counter, DRAM RAS and CAS generation, video control signals, memory decode, CPU clock generation, contention handling and most of the pixel shift register.

The ULA area used does not extend to the outer edge of matrix cells, with the exception of DL4, AL4 and PL4, as the design was originally laid out for a 5C000 series ULA that contained 11 × 10 matrix cells per quarter instead of 12 × 11 for the 6C000 series.

Running around the outside of the ULA, surrounding the matrix cells and crossing over the peripheral cells, are the four ULA power rails. The outer thin rail is GND, and inside that is the 5v peripheral cell I/O supply. The next and thickest power rail is the logic 5v supply, which is regulated into a 0.95v matrix cell supply by serial voltage regulators located at the base of each peripheral cell. These regulators are controlled by a bandgap reference voltage rail, Vref, that runs inside the logic rail, producing the innermost matrix cell Vs rail.

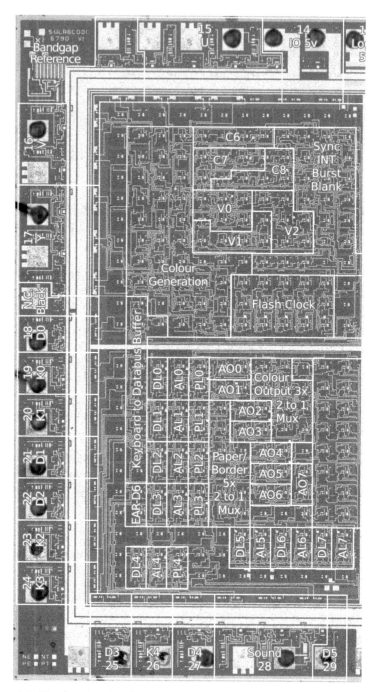

*Figure A-1: Die plot of the ZX Spectrum 6C001 ULA - Quarters 1 and 3*

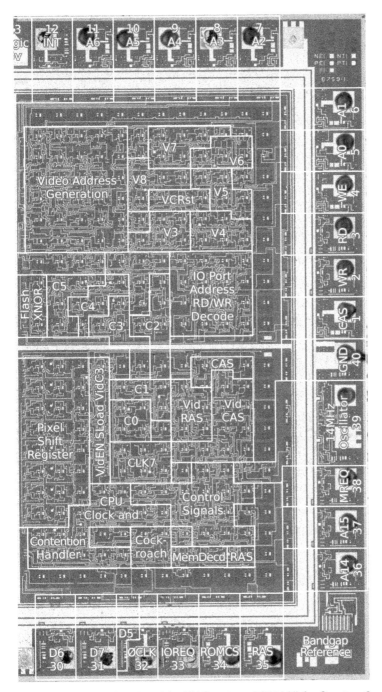

*Figure A-2: Die plot of the ZX Spectrum 6C001 ULA - Quarters 2 and 4*

| Identifier | Description |
|---|---|
| AL7-0 | Attribute byte data latch consisting of eight gated D transparent latches. |
| AO7-0 | Attribute output latch consisting of eight gated D transparent latches. |
| C5-0 | The master counter D-type flip-flops. |
| C8-6 | The master counter T-type flip-flops. |
| CAS | CPU DRAM CAS generation. |
| CLK7 | Divides the 14 MHz master clock into a 7 MHz signal with an even duty cycle. Drives the pixel shift register and master counter. |
| Cockroach | The issue 1 "dead cockroach" fix when implemented within the 5C112 and 6C001 ULAs. |
| Colour Generation | Logic to generate the various RGB, timing, blank and black signals required for the analogue luminance and chrominance signal generation. |
| Colour Output Mux | The colour output multiplexer from the attribute output latch. |
| Control Signals | The video generation and memory fetch control and timing signals. |
| CPU Clock and Contention Handler | Generation of the CPU clock and contention handling. |
| DL7-0 | Display byte data latch consisting of eight gated D transparent latches. |
| EAR-D6 | IO port output buffer to CPU data bus of EAR socket state. |
| Flash Clock | The 5-bit flash ripple counter consisting of D-type flip-flops. |
| Flash XNOR | XNOR of serial data stream from pixel shift register with the flash clock. |
| IO Port Address RD/WR Decode | IO port address decoding and test mode decoding. |
| Keyboard to Data Bus | IO port output buffer to CPU data bus of keyboard state. |
| MemDecd | CPU memory access decoding and generation of /ROMCS and RAM16 signals. |

| Identifier | Description |
|---|---|
| Paper / Border Mux | Multiplex the background and border colours into the attribute output latch. |
| Pixel Shift Register | The pixel shift register, constructed from eight D-type flip-flops with input multiplexing. |
| PL4-0 | IO port output latch consisting of five gated D transparent latches. |
| RAS | CPU DRAM RAS generation. |
| Sync, INT, Burst, Blank | Counter driven logic that derives the video synchronisation, burst and blank signals as well as the CPU interrupt request signal. |
| V9-0 | The master counter T-type flip-flops. |
| VCRst | Vertical counter reset logic. |
| VidEN, SLoad and VidC3 | Further specific control signal generation. |
| VidCAS | Video access DRAM CAS generation. |
| VidRAS | Video access DRAM RAS generation. |

# Appendix B
# Component Library

The Ferranti component library for each type of ULA array contained a comprehensive collection of functional building blocks that designers could draw upon. Each unit was provided with details of relevant characteristics, such as propagation delay and fan-out, and in the case of interface functions, the current and voltage levels.

In addition to the circuit schematic, example interconnect routing patterns are occasionally given, an example of which is presented in Figure B-13.

## Flip-Flops and Latches

The flip-flops and latches used by the ZX Spectrum ULA are presented in this section. The flip-flops and shift registers are negative edge triggered, and the latches enter a latched state when the enable input is high.

| Description | Reference | Cells | Figure |
|---|---|---|---|
| T-Type with reset, carry, enable | TRCE | 5 | Figure B-1 |
| T-Type with carry, enable | TCE | 5 | Figure B-2 |
| T-Type with reset, carry | TRC | 3.5 | Figure B-3 |
| D-Type flip-flop | FD | 3 | Figure B-4 |
| Single bit of shift register | SHIFT | 5 | Figure B-5 |
| 8-bit shift register | Shift8 | 40 | Figure B-6 |
| Gated D transparent latch | GD | 2 | Figure B-7 |

*Table B-1: Matrix cell counts of the latches and flip-flops used by the ZX Spectrum*

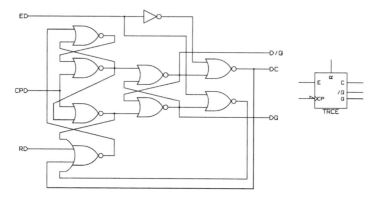

*Figure B-1: T-Type flip-flop with Reset, Carry and Enable*

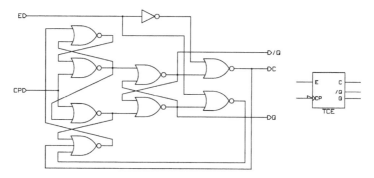

*Figure B-2: T-Type flip-flop with Carry and Enable*

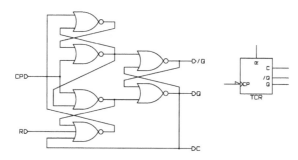

*Figure B-3: T-Type flip-flop with Reset and Carry*

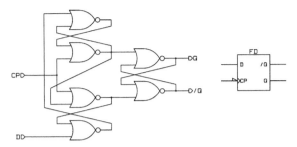

*Figure B-4: D-Type flip-flop*

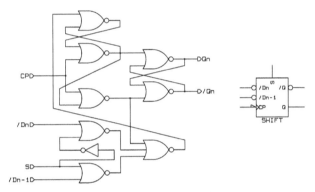

*Figure B-5: Single bit stage of shift register*

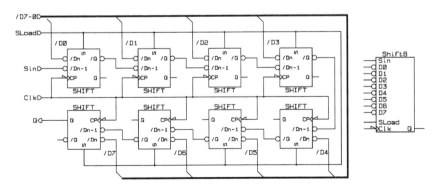

*Figure B-6: 8-bit shift left register*

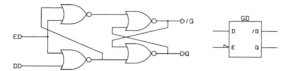

*Figure B-7: Gated D transparent latch*

## Interface Functions

Input and output functions, such as the TTL compatible inputs used by the ZX Spectrum ULA are presented in this section.

The resistor values given are for the 6000 series ULA peripheral cell, such as the 6C001E-6 and 6C001E-7. Figure B-13 shows an example interconnect layout for the tri-state totem pole output given in Figure B-12.

| Description | Direction | Figure |
|---|---|---|
| Open-collector output with high fan-out | Output | Figure B-8 |
| TTL compatible open-collector output, emitter-follower input | Bidirectional | Figure B-9 |
| TTL compatible input | Input | Figure B-10 |
| Emitter-follower buffer input | Input | Figure B-11 |
| Totem pole output with tri-state capability | Output | Figure B-12 |

*Table B-2: Interface functions used by the ZX Spectrum*

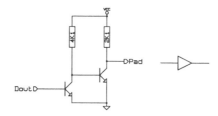

*Figure B-8: Open-collector output with strong pull-up*

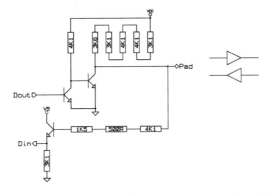

Figure B-9: TTL compatible open-collector output and emitter-follower input

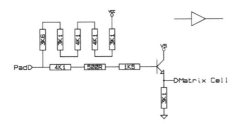

Figure B-10: TTL compatible input

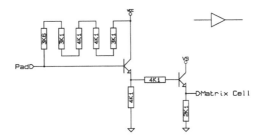

Figure B-11: Emitter-follower buffer inputs

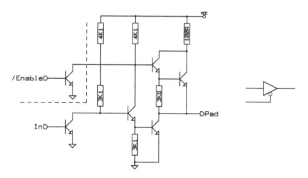

*Figure B-12: Totem pole output with tri-state capability*

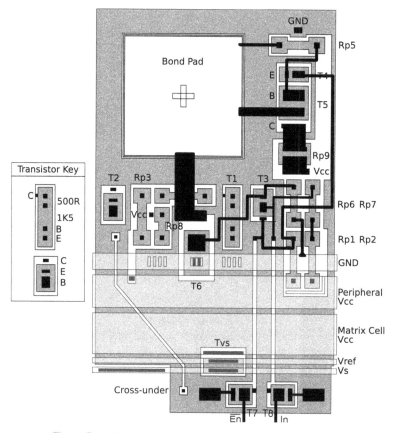

*Figure B-13: Tri-state totem pole peripheral cell interconnect*

# Appendix C
# ULA Configuration

The basic gate type supported by the ULA matrix cell is the 2-input NOR gate. A single 5000 and 6000 series matrix cell will provide up to two 2-input NOR gates. The propagation delay and fan-out of these gates is determined by the load resistor and current source configuration they have, in addition to the number of inputs they have (fan-in).

The standard 2-input NOR gate is shown in Figure C-1. Here, all four transistors of the matrix cell are used to provide two identical 2-input NOR gates, each with a single load resistor, RL, and current source connection.

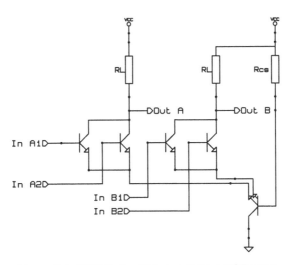

*Figure C-1: 6000 Series ULA, two 2-input NOR gates*

These two 2-input NOR gates may be combined into a single 4-input NOR gate by joining the transistor collectors together, along with the emitters. See

Figure C-2. As twice the number of transistors are being used for a single gate, to maintain the same propagation delay and fan-out capability, it is important that the load resistor be halved in value, and the current source doubled. Hence the two available load resistors are connected in parallel, and both current sources are connected to the transistor emitter.

If the gate were required to switch slower, then only one load resistor and current source would be used.

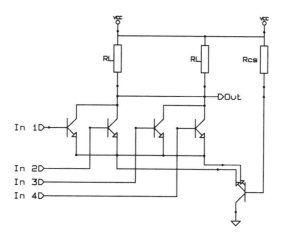

*Figure C-2: 6000 Series ULA, 4-input NOR gate*

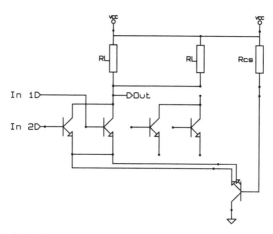

*Figure C-3: 6000 Series ULA, 2-input NOR gate, reduced propagation delay*

The technique of increasing the current drawn through the gate transistors can be applied to any gate, decreasing its propagation delay and increasing its fan-out capability. Figure C-3 shows a 2-input NOR gate, configured with two parallel load resistors and two coupled current sources. This configuration would approximately reduce the propagation delay of the gate by one quater.

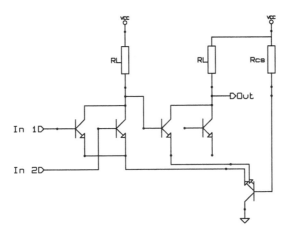

*Figure C-4: 6000 Series ULA, 2-input OR gate*

Large gates, such as the 8-input NOR gate shown in Figure C-5, are constructed by connecting the transistors from multiple matrix cells in parallel. Again, the appropriate number of load resistors and current sources are used to configure the required propagation delay and fan-out.

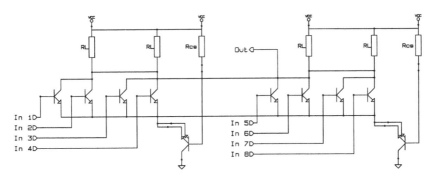

*Figure C-5: 6000 Series ULA, 8-input NOR gate*

Figure C-6 shows an example matrix cell interconnect for a D-type flip-flop. Note that the /Q gate is configured with a load resistance of RL/2 and two current sources, giving it a higher fan-out that the Q output. This example is taken from the C0 master counter stage of the ZX Spectrum 6C001 ULA.

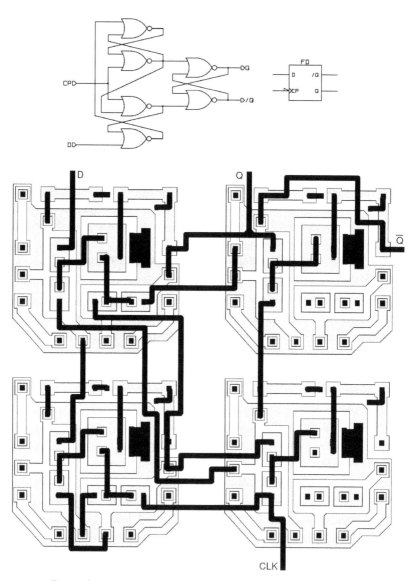

*Figure C-6: D-Type flip-flop using three out of four matrix cells*

# Appendix D
# Mathematical Proofs

## Quadrature Amplitude Modulation

The following proof demonstrates that the sum of two orthogonal, amplitude modulated carriers results in an output that is both phase and amplitude modulated.

$$f(x) = I \sin(2\pi f_c t) + Q \cos(2\pi f_c t)$$

$$x = 2\pi f_c t$$

Quadrature modulation, defined by:

$$f(x) = a \sin(x) + b \cos(x)$$

$$f(x) = \sqrt{a^2 + b^2} \left( \frac{a}{\sqrt{a^2 + b^2}} \sin(x) + \frac{b}{\sqrt{a^2 + b^2}} \cos(x) \right)$$

From trigonometry:

$$\cos(\alpha) = \frac{a}{\sqrt{a^2 + b^2}}$$

$$\sin(\alpha) = \frac{b}{\sqrt{a^2 + b^2}}$$

substituting:

$$f(x) = \sqrt{a^2 + b^2} (\cos(\alpha) \sin(x) + \sin(\alpha) \cos(x))$$

$$f(x) = \sqrt{a^2 + b^2} (\sin(x + \alpha))$$

$$f(x) = \sqrt{I^2 + Q^2} (\sin(2\pi f_c t + \alpha))$$

# Bibliography

[ALTWASSERDC] Richard Altwasser, *Display for a computer*, 10th November 1983, Sinclair Research Limited, Richard Altwasser, Patent WO 1983/003916, Sinclair Research LTD, 9th May 1984, Patent EP0107687A1, Amstrad PLC, 6th July 1988, Patent EP0107687B1.

[ALTWASSERIO] Richard Altwasser, *Computer input/output circuit*, 21st April 1983, Sinclair Research Limited, Patent GB2119207.

[BRUCH] W Bruch, *The PAL colour TV system: basic principles of modulation and demodulation*, 1964, NTZ Communications Journal, 255-268.

[CPMAN1970] The Conservative Party, *Conservative Party general election manifesto – A better tomorrow*, 1970.

[DS4116] NEC Electronics, *μPD416 16,384 × 1-Bit Dynamic NMOS RAM*, 1984, 416DS-REV4-1-84-CAT-L.

[EDNDC] Ian Hickman and Bill Travis, Butterworth Heinemann, *EDN designers companion*, 02/09/1994, 978-0-750617-21-5, Annotated Edition.

[FERRANTIRS] Ferranti Electronics Limited, *Technical brief 2 – 'R' series ULA*, 2.001, Nov 1982, National Archive 1996.10/7/7/33.

[FERRANTILSI] Ferranti Semiconductors Limited, *Ferranti leaders in LSI array technology*, 1979, National Archive 1996.10/7/6/86.

[FERRANTIPG1] Ferranti Semiconductors Limited, *ULA product guide*, 1982, National Archive 1996.10/7/7/33.

[FERRANTISG1] Ferranti Semiconductors Limited, *Ferranti ULA selection guide*, 1979, National Archive 1996.10/7/6/86.

[FERRANTIUDB] Ferranti Semiconductors Limited, *Ferranti ULA Designer system backgrounder*, 23/2/1982, National Archive 1996.10/7/9/12.

[NOYCE] Robert N. Noyce, *Semiconductor device-and-lead structure*, 30th July 1959, U.S. Patent 2,981,877.

[HOERNI] Jean Horeni, *Method of manufacturing semiconductor devices*, 1st May 1959, U.S. Patent 3,025,589.

[HURSTCSIC] Stanley L. Hurst, Marcel Dekker, Inc, *Custom-specific integrated circuits*, 1985, 0-8247-7302-0.

[HURSTVLSI] Stanley L. Hurst, Marcel Dekker, Inc, *VLSI custom microelectronics*, 1999, 0-8247-0220-4.

[LOGAN] Dr Ian Logan and Dr Frank O'Hara, Melbourne House (Publishers) Ltd, *The complete Spectrum ROM disassembly*, 1983, 0-8616-1116-0.

[MICROMATRIX] Cloyde E. Marvin and Robert M. Walker, Electronics, *Customizing by interconnection*, Feb 20th, 1967.

[MURPHY] B.T. Murphy, V.J. Glinski, P.A. Gary, and R.A. Pedersen, IEEE, *Collector diffusion isolated integrated circuits*, 1969, IEEE Volume 57, Issue 9. Pages 1523 – 1527.

[RAMSAYAUTO] Frank R. Ramsay, Ferranti Electronics Limited, *Automation of design for Uncommited Logic Array*, 1980, ACM 0-89791-020-6/80/0600/0100.

[RAMSAYREM] Frank R. Ramsay, Ferranti Electronics Limited, *A remote design station for customer Uncommited Logic Array designs*, 18th Design Automation Conference, 1980, IEEE 0146-7123/81/0000-0498.

[SM48] Thorn (EMI) Datatech Ltd, Sinclair Research Ltd, *Sinclair ZX Spectrum Service Manual*, March 1984.

[SWANN] Peter Swann and Jasvinder Gill, Routledge, *Corporate vision and rapid technological change: The evolution of market structure*, 1993, 0-4150-9135-7, 978-0-415091-35-0.

[TOZER] EPJ Tozer, *Broadcast engineer's reference book*, 14th May 2004, Focal Press, 978-0-240519-08-1.

[ULAHAND] Ferranti Electronics Limited, *The ULA technical handbook*, Issue 1, Nov 1980, National Archive 1996.10/7/6/86.

[VALENSI] Georges Valensi, *System of television in colors*, 15th May 1945, U.S. Patent 2,375,966.

[VIDEODM] Keith Jack, Newnes, *Video demystified: A handbook for the digital engineer*, 1st May 2007, 978-0-750683-95-1, 5th Edition.

[WILSONTT] John F. Wilson, ICBTT, *Technology transfer and the british microelectronics industry, 1950–75: Confused signals*, 2004, http://www.jsme.or.jp/tsd/ICBTT/conference02/JohnWILSON.html.

[WILSON2] John F. Wilson, Crucible Books, *Ferranti: A history*, Volume 2, 2007, 978-1-905472-01-7.

[Z80UM] Zilog Inc, *Z80 Family CPU User Manual*, December 2005, UM008005-0205 Revision 05.

[Z80ZDS] Zilog Inc, *Z80400/Z84C00 NMOS/CMOS Z80 CPU Central Processing Unit*.

[Z80SDS] Sharp, *LH0080 Z80 CPU Central Processing Unit*.

# Glossary

Glossary of terms used in the book.

## Analogue

Electronic functionality that produces or processes signals that vary over a range of voltages.

*See Also:* Linear, Digital.

## ADC

Analogue to digital converter. Take a variable voltage and converts it to a binary representation that can be processed and stored by the computer.

*See Also:* DAC.

## Backcloth

The term "backcloth" refers to the background template displayed by a computer aided design (CAD) application. These aid the placement of components and lines, the simplest example of which is a grid.

## Bit

A single binary digit and is represented by the values 0 or 1.

## Byte

A byte is comprised of eight bits, representing a decimal value between 0 and 255.

## CDI

Collector Diffusion Isolation.

## Complete Gate

A Complete Gate is one from which all other logic functions may be derived. NAND and NOR gate provide such function.

## CML

Current Mode Logic.

## CPLD

Complex Programmable Logic Device. An advanced type of programmable gate array based around a macrocell design. They offer greater functional complexity than the earlier Programmable Logic Array (PLA), but less than the Field Programmable Gate Array (FPGA).

*See Also:* FPGA.

## DAC

Digital to analogue converter. Take a binary representation of a voltage level from the computer and produces an equivalent output voltage.

*See Also:* ADC.

## Digital

Electronic functionality that produce or processes signals that consist of two voltage levels: on or off.

*See Also:* Analogue, Linear.

## Doping

The addition of impurities to silicon to change its physical properties. Doping with periodic table group 13 element such as boron produces p-type silicon. Doping with a group 15 element such as arsenic produces n-type silicon.

*See Also:* N-Type, P-Type.

## DTL

Diode Transistor Logic.

## Distributive Law

The Distributive Law of boolean algebra defines the following relationships:

$$A \cdot (B + C) = (A \cdot B) + (A \cdot C)$$
$$A + (B \cdot C) = (A + B) \cdot (A + C)$$

## Epitaxial Layer

A layer of silicon that has been grown on top of another layer by using the Epitaxial Process.

*See Also:* Epitaxial Process.

## Epitaxial Process

A technique for growing a layer of silicon on top of another by using a hot chemical vapour.

## FPGA

Field-Programmable Gate Array.

A field-programmable gate array is a gate array that is designed to be programmed within the end-user equipment. It does not store its programming, which must be supplied from an external memory device such as an EEPROM on power-up.

## IC

*See:* Integrated Circuit

## Integrated Circuit

A multi-transistor circuit implemented on a small piece of silicon. It is more correctly referred to as the monolithic integrated circuit, and is also known as the silicon chip due to its size and appearance.

## Light Field Mask

A light field mask contains isolated dark mask regions within and surrounded by a radiation transmitting background.

*See Also:* Photolithography, Photoresist.


## Linear

Electronic functionality that produces or processes signals that vary over a range of voltages.

*See Also:* Analogue, Digital.


## LSI

Large scale integration

*See Also:* SSI, VLSI.


## M-cycle

Operations within the Z80 are called M (machine) cycles. An M-cycle consists of four, five or six T-states. The first M-cycle, M1, of an instruction is the instruction fetch cycle, followed by as many memory or I/O read and write cycles necessary to complete the instruction.

*See Also:* T-state.


## Micro-

Units of $10^{-6}$. Given the symbol '$\mu$', such that, for example, microseconds = $\mu$s, milliamp = $\mu$A.

$5 \ \mu s = 5 \times 10^{-6} \ s = 0.000005 \ s.$

*See Also:* Milli-, Nano-.

## Milli-

Units of $10^{-3}$. Given the symbol 'm', such that, for example, milliseconds = ms, milliamp = mA.

$5 \text{ ms} = 5 \times 10^{-3} \text{ s} = 0.005 \text{ s}$.

*See Also:* Micro-, Nano-.

## Multiplexer

A logic circuit that switches an output between multiple inputs. A 2-to-1 multiplexer switches one of two possible inputs to a single output, a 4-to-1 multiplexer switches one of four inputs to a single output.

## Nano-

Units of $10^{-9}$. Given the symbol 'n', such that, for example, nanoseconds = ns, nanofarads = nF.

$5 \text{ ns} = 5 \times 10^{-9} \text{ s} = 0.000000005 \text{ s}$.

*See Also:* Milli-, Micro-.

## NPN

A transistor formed by sandwiching a p-type silicon layer between two n-type layers.

*See Also:* PNP, P-Type, N-Type.

## NTSC

National Television System Committee. The analogue television encoding standard used primarily in the United States.

## N-Type

Silicon that has been doped with a periodic table group 15 element such as arsenic.

*See Also:* P-Type, Doping.

## OEM

Original equipment manufacturer. For example, Sinclair Research.

## Ohm's Law

Ohm's Law states that at a constant temperature, the current flowing through a conductor (I) is directly proportional to the potential difference across the conductor (V), and inversely proportional to the conductors resistance (R).

$$I = \frac{V}{R}$$

## Op Code

The operation code that specifies the operation to be carried out. Also referred to as instruction code.

## Open Collector

An open collector output is one in which is taken directly from the collector of an output transistor whose emitter is tied to 0v. Applying a voltage to the base of the output transistor pulls the collector down to 0v. When no voltage is applied to the base, the collector output is effectively disconnected, or open-circuit.

Open collector outputs are common in integrated circuits.

## Orthogonal

Signals, usually sine waves, that are 90 degrees out of phase with one another. Sine and cosine signals are orthogonal.

## Oxide Mask

A template formed from silicon dioxide and laid over the surface of the integrated circuit during manufacture. Silicon is then applied on top of the mask before being removed, leaving silicon deposited in the desired areas.

## PAL

Phase-alternate line. The standard for analogue video encoding used primarily in the United Kingdom.

## PCB

A PCB or *Printed Circuit Board* is the sheet of fiberglass onto which electronic components are mounted. These components are interconnected by tracks of copper.

## Photolithography

The technique of laying down a light sensitive coating, called a *photore-*

*sist*, which is selectively hardened by exposure to a pattern of light. The unexposed areas of coating are then chemically etched away to leave a mask template over the surface.

*See Also:* Photoresist, Light Field Mask.

## Photoresist

A light sensitive coating that comes in two forms: a positive photoresist that allows the exposed areas to be etched away, and a negative photoresist that allows the unexposed areas to be etched away.

*See Also:* Photolithography, Light Field Mask.

## Pinch Resistor

A pinch resistor is an area of p-type silicon sandwiched between two horizontal layers of n-type silicon. In the CDI process, such a resistor would be formed as a normal diffused resistor of p-type silicon, which has had a shallow n+ type silicon region diffused over the top, between the resistor contacts. This 'pinches' the resistor vertically and increased the effective resistance.

## Planar Process

A method of fabricating silicon circuits by laying down successive layers of silicon over a silicon substrate, forming interconnected transistors and other components.

## PNP

A transistor formed by sandwiching an n-type silicon layer between two p-type layers.

*See Also:* NPN, P-Type, N-Type.

## Propagation Delay

Propagation delay is the time taken for a logic gates output to change in response to an input. Several factors may increase the propagation delay of a gate, including having an excessive load connected to its output or a reduced supply voltage.

## P-Type

Silicon that has been doped with a periodic table group 13 element such as boron.

*See Also:* N-Type, Doping.

## Pull-Up

The act of raising the voltage of a signal so that it represents a logic 1. This is usually achieved through a "pull-up" resistor that is connected between the signal and Vcc.

## Register

1. A latch within a circuit or IC that may be written to by the CPU to control its behaviour.

2. One of many data latches within a CPU used for temporary information storage during the execution of Op Codes.

*See Also:* Op Code.

## RF

Radio Frequency.

## RTL

Resistor Transistor Logic.

## Reticle

A grid or pattern placed in the light path of an optical instrument.

## SBC

Standard Buried Collector is a term applied to a particular form of bipolar transistor construction, since it requires a buried layer of n+ silicon below the transistor fabrication, within the P-type substrate.

## SECAM

Séquentiel Couleur à Mémoire. Also known as Sequential Colour with Memory. The analogue video encoding standard used primarily in France.

## Serialise

The process of turning a byte or similar group of bits into a sequential binary stream. Usually achieved with a shift register that is simultaneously loaded with all bits in parallel, each of which is output in turn through a single channel with each cycle of the clock.

## SPICE

Simulation Program with Integrated Circuit Emphasis.

SPICE is a program that allows you to simulate electronic circuits with a computer. It can calculate voltages and currents with respect to time, called transient analysis, or frequency, called AC analysis.

## Square Wave

A square wave signal is one where the transition between states occurs very rapidly. Square waves are common in digital circuits as they permit logic to be switched at precise moments in time.

## SSI

Small scale integration

*See Also:* LSI, VLSI.

## Substrate

The rigid, thin and perfectly flat silicon wafer on which integrated circuits are manufactured.

## Suppressed Carrier Quadrature Modulation

The amplitude of two carrier signals, 90 degrees out-of-phase with one another (in quadrature), are modified in response to two data signals, and then summed. The signal that results consists of a single carrier that is both amplitude and phase modulated by the two data signals. Being amplitude modulated, the carrier signal is suppressed.

## Test Vector

A test vector is a set of input conditions and associated output conditions used to test a hardware design.

## Tri-State

Tri-state refers to an integrated circuit logic output that can have three states, defined as: logic 0, logic 1 or high impedance (effectively disconnected).

## TTL

Transistor-transistor logic.

TTL ICs are five volt devices whose logic output swings from less than 0.35v for a logic 0, to between 3.3 and 5v for a logic 1. For these devices to correctly register a logic input, the signal must be less than 0.8v for a logic 0 and greater than 2v for a logic 1.

Logic families 74xx, 74LSxx, 74Fxx, 74ASxx and 74ALSxxx are all TTL devices, the 74LS being the most common in the 1980s.

## T-state

A T (time) state is the smallest unit of time taken by any Z80 operation and is equal to one clock cycle. Operations take between four and six T-states to complete, where the first T-state is referred to as T1, the third as T3, and so on.

*See Also:* M-cycle.

## UV

Ultraviolet light

## VLSI

Very large scale integration

*See Also:* LSI, SSI.

# Index

# Colophon

## Colophon

This book was created on a GNU/Linux workstation, running Debian Squeeze/Sid. The source text was written in Docbook V4.5 XML using vi and XMLMind XML Editor for spell checking, and processed using openjade and Norman Walsh's DocBook DSSSL stylesheets, which were overridden to produce the desired layout. Each chapter was produced as a separate XML file and preprocessed by xmllint into a single document prior to passing to openjade.

Diagrams were produced using inkscape, imported into GIMP for resizing and to allow touch up before being exported as PNG files. Schematics were drawn using kicad.

The index was created by preprocessing the document for SGML output through openjade with the html-index option, and then collating the HTML index with Norman Walsh's collateindex.pl to produce an XML Docbook index file.

Math equations were written in TeX, escaped and embedded in the docbook XML. The openjade TeX output was processed to unescape the math TeX prior to rendering the final output.

The cover features an issue 3B ZX Spectrum 48K PCB, back lit by a high intensity red lamp. The photograph was distorted into a sphere using Gimp, separated into CMYK plates using the Separate+ plugin and exported as a CMYK tif. The cover itself was laid out using Scribus, from which it was exported as a print ready PDF.

The main text of the book is set in Nimbus Roman. Titles and headings are set in Nimbus Sans.

Colophon

www.ingramcontent.com/pod-product-compliance
Lightning Source LLC
Chambersburg PA
CBHW071406050326
40689CB00010B/1769